Springer Geography

The Springer Geography series seeks to publish a broad portfolio of scientific books, aiming at researchers, students, and everyone interested in geographical research. The series includes peer-reviewed monographs, edited volumes, textbooks, and conference proceedings. It covers the entire research area of geography including, but not limited to, Economic Geography, Physical Geography, Quantitative Geography, and Regional/Urban Planning.

More information about this series at http://www.springer.com/series/10180

Emilie Lavie · Anaïs Marshall
Editors

Oases and Globalisation

Ruptures and Continuities

Springer

Editors
Emilie Lavie
Université Paris-Diderot
Paris
France

Anaïs Marshall
UFR LLSHS, Département de Géographie
Université Paris 13-Nord
Villetaneuse
France

ISSN 2194-315X ISSN 2194-3168 (electronic)
Springer Geography
ISBN 978-3-319-50747-7 ISBN 978-3-319-50749-1 (eBook)
DOI 10.1007/978-3-319-50749-1

Library of Congress Control Number: 2016961304

Printed on acid-free paper

This Springer imprint is published by Springer Nature
The registered company is Springer International Publishing AG
The registered company address is: Gewerbestrasse 11, 6330 Cham, Switzerland

Foreword

Novelists, painters, geographers, and travellers in general have long contributed to making oases a lingering symbol of orientalist aesthetics. From the image of a palm tree grove feeding on a spring in the middle of sand dunes, many other often isolated spots of greenery associated with a harsh surrounding environment, that precludes or constrains settlement and agricultural activities, have come to be termed or considered as "oases". The collection of case studies presented in this book widens our imagination as well as our understanding of the specificities of such unique and diverse locales.

Isolation, though, is largely a myth, as the book demonstrates, not only because some oases—close to the coast, like in Peru, or located on major commercial roads, like in Xinkiang—have long been booming commercial crossroads and even kingdoms (Palmyre or Thar), but also because the oases that were hitherto relatively insulated by their environment are the object of—sometimes brutal—drivers of changes. The collection of papers shows that drivers may be both endogenous—when the delicate balance between humans and their resources is disrupted—or exogenous, when the global forces and logics of capital accumulation swoop down on the economic potential offered by a peculiar combination of soil and water (and sun) resources.

These new investors may be residing in larger nearby towns and willing to keep a link with their birthplace, outsiders investing in agricultural, tourism or other ventures, larger companies or even land developers (with a varying mix of national and transnational capital), or the state that pushes particular projects. Roads open them up and expose oases to the outer world and global markets; tourists flow in with money and other cultural values; truck drivers make stopovers and new crops are established; state administrations expand.

Newcomers like agribusiness investors are linked to global markets (and often to national politics) and the multiple interactions and competitions generated by their intrusion expand the chain of causalities of local transformations. These competitions may concern land, labour, services or even agricultural production itself, for example, but are generally more saliently illustrated by the issue of water allocation and management. A primary disruption of the subtle balance between oasis societies and their environment is a gradual or brutal leap in the amount of water abstracted from the basin, with a wide range of configurations in the respective importance of surface and groundwater and their interrelationships, as well as in the ensuing spatial/social redistribution of costs and benefits. Although several cases presented in the book show that oases should not be seen as passive victims of a forced integration into a wider globalised word, pointing to the agency deployed by local actors, many transformations nevertheless come with increased social differentiation within oasis systems. While they may rise from a hamlet to regional prominence, they can also regress from supremacy to marginalisation, swallowed up by bigger cities, eaten away by salt accumulation, withered by vanishing waters, or even become a mere water storage area for megalopolis, like Liwa in the UAE!

Environmental and cultural heritage values have also entered the scene, mobilising the folk image of an oasis as a lush haven of peace and greenery surrounded by an adverse environment, now promoted as a symbol of the crave of urbanites for places insulated from the hustle and bustle of everyday life and for the "consumption" of an exotic nature.

Like with river deltas, floodplains or small islands, geographers have long been attracted by the peculiarities of oases, their often tightly knit societies and the social rules they crafted for the capture and distribution of water, their delicate balance between resources and population, a prime instance of what nowadays would be termed socio-ecological systems. They, together with the wider public, will certainly enjoy this book and read it as a fascinating travelogue through various continents and landscapes which, eventually, all hang by an often thin thread of water.

François Molle
IRD, UMR GRED
Montpellier, France

Contents

Introduction

"Oasis"… when we read or hear this word, each one of us imagines a particular landscape, mainly consisting of water and palm trees surrounded by a desert of sand, or else a haven, an Eden, a place to rest. The oasis is often perceived as a place in contrast with its environment. Moreover, the many accounts of travellers and explorers, as well as the reports of scientists up to the beginning of the twentieth century, have promoted this fixed image of the oasis in opposition to the hostility of the surrounding area: a fragile cultivated space at the heart of a hostile arid area. For a long time, geographers also kept in their descriptions this idea of opposition between oasis and desert, coupled with the hostility of nature: "*The oasis can be defined by the effect of the contrast between its island of dense greenery and the arid or semi-arid areas that surround it*" (Mainguet 2003: 240); "*the antithesis of desert*" (George, quoted in Mainguet 2003: 240). Garcier and Bravard (2013 and 2014) revisit the origin of this spatially and temporally fixed vision of the oasis, both in its internal organisation and in its existence. These authors explain that the "oasian model" of the Libyan Desert, a product of colonial development, assumed that the natural environment was stable and that "*the variations observed in the environment* [were due to] *the unequal competence of civilisations and human action against the forces of nature*". In this way, many studies focus on water resources and their accessibility, control and management by societies. Water is thus considered central in the development and maintenance of oases. However, as some authors state (Lacoste 1990; Garcier and Bravard 2013), especially Battesti (2005: 12), "*water is a necessary but insufficient condition to explain the creation of an oasis*". In fact, the formation and maintenance of oases are the result of several historical, political and social factors. Historically, oases are often ancient inherited settlements of nomadic territories. In Africa and the Middle East, they were developed due to their role as a stopping-place or caravanserai. An analysis of the history of places is thus fundamental in order to understand current realities. Politically, as Garcier and Bravard (2013: 3) describe, "*there is no oasis without intention, or without work, which often takes the form of organised political projects*". It is thus also essential to take into account political decisions and planning and development strategies when analysing the evolution of oases. Lastly, socially, the oasis is a territory in which the actors adapt to a constantly changing environment by shaping and creating a particular oasian landscape. Indeed, although each oasis is specific, there are certain similarities in their territorial organisation and in their spatial and temporal evolution.

It was these evolutions, studied through the lens of globalisation, that stimulated the organisation of a symposium "Oases in globalisation: ruptures and continuities" in December 2013 in Paris. This brought together researchers and specialists of various oases in order to look at their characteristics and realities in the context of globalisation. In this approach, these oasian areas are re-examined in terms of recent spatial and temporal dynamics. Have some oases disappeared? Have others been created? Can we still use the term "oasis" for these changed areas? What about the new irrigation schemes created from scratch in arid areas?

However, before discussing these dynamics and concepts, we should first return to the process of globalisation. Since the beginning of the 1990s, "globalisation" has described a new phase in the worldwide integration of economic, financial, ecological and cultural phenomena. Exchanges are growing on a global scale, and the process of globalisation is resulting in a multitude of all types of changes and a worldwide geopolitical reconfiguration (Dollfus 1997; Lombard et al. 2006). The study of globalisation often covers its economic and financial aspects but should also include the opening up of markets, especially in property, the social, political, environmental and territorial dynamics, and their impacts. In addition, regional differences must be taken into account, given that the perception of globalisation is not the same in the "Northern" countries as in the "Southern" countries (Gervais-Lambony and Landy 2007). Moreover, Lombard et al. (2006) point out that the special feature of the "Southern" countries is "*the speed and the violence with which these changes occur*". The spaces studied in this book all belong to the countries of the South in particular for the rapid and dynamic changes.

Many oases joined the market economy (Clouet 1995) very early on at the global scale and have played an important strategic role in arid areas. In a special edition on *Islands and Oases*, Troin (2005: 339-341) wrote that he saw oases as islands in the desert, "*The resulting isolation due to the hostile environment, the distance, the climatic conditions, the difficult access, the small population, and the fragility of resources constituted a topic widely studied by our predecessors. (…) In short, the isolation was rather the other face of networks, the hole between the mesh, it created the isolate. But the isolate is no longer what it was. (…) Every inhabited place on Earth is permanently accessible (…), openness to the world is global, places that used to be isolated, fortified, and coveted have become crossroads, an escape for those bored with civilisation, or even "global" shopping centres*".

In the current process of globalisation, the oasis is undergoing significant changes, both at the scale of the oasis and its components and at the scale of the region where it is located. These changes may be political (upheavals like decolonisation, the cultural revolution in western China at the end of the 1960s or, more recently, the Arab spring), economic (particularly the expansion of irrigation with agro-industrial crops destined for export, and global competition that challenges the very existence of an oasis or a part of oases), social (such as oasian emigrations or the arrival of new actors and the resulting exchanges), cultural (especially the designation of oases as heritage areas for tourism), environmental (at the local, regional or global scale) and/or territorial (urban growth and its pressure on property, for example).

Following the symposium of the same name, this book *Oases in Globalisation: Ruptures and Continuities* has mainly been written by geographers (but also by archaeologists, literature researchers, sociologist and agronomist) with the goal of analysing these changes, the landscapes and the intrinsic and regional structures of some oases, in a context of openness and increasing and varied exchanges on a global scale.

Part I entitled "Oasis?" has a certain epistemological objective. In fact, our research on this term revealed almost exclusively French and Chinese references[1] on the concept of an oasis, which were often quite old. Beyond the classic "island of greenery in the desert", the interest in discussing the term oasis is to suggest a new definition, based on current realities. Do the local and regional dynamics challenge the use of the term "oasis" to define these areas? Bénédicte Veyrac Ben-Ahmed and Slaheddine Abdedayem (Chapter "Oases in Southern Tunisia: The End or the Renewal of a Clever Human Invention?") raise the question of defining an oasis, given that there are so many different types. The oases, by losing some of their functions or simplifying their systems, have also lost what makes them unique. Does the concept need to be updated? They try to answer this question using the example of the Tunisian oasis of Gabès. Marc Kober, a researcher in Literature (Chapter "A Travel Through Oases in French and Arabic Literature"), provides an interpretation of the oasis in the French colonial literature from the turn of the twentieth century, in which the (real) Saharan oases are the backdrop to a "Robinsonade", and in contemporary Middle East literature, in which imaginary oases give a mystic significance to the adventure of their heroes. Lastly, Evelyne Mesclier, Anaïs Marshall, Célia Auquier and Jean-Louis Chaléard (Chapter "The Conceptual Approach of Oasis as Insights on Globalisation. Example of the Coastal Valleys of Northern and Central Peru.") propose a reading of the valleys and interfluves of the Peruvian coast by the oasis concept. The approach is not focused on the landscape but from the systemic spatial relationships. This entry allows them to analyse local and spatial impacts of globalisation through the case of agribusiness companies that are reshaping all territorial relations of irrigated areas.

Part II "History of Water and Oases" gives a historical perspective of the changes in oasian landscapes in relation to irrigation. Is water really the control valve of the creation and development of oases? Nothing is less certain. In the context of climate change and the desertification of arid zones, water may be lacking. Yet, on one hand, oasian populations are particularly resilient to a varying water supply; on the other hand, many authors have deconstructed the principle of shortage, which may be

[1]The Xinjiang Institute of Ecology and Geography of the Chinese Academy of Sciences in Urumqi regularly publishes articles on the nature of oases, but they are not very accessible from abroad, and only written in Chinese.

more built than natural, often justifying water projects (Postel 1992; Aguilera-Klink et al. 2000; Bakker 2000; Metha 2001; Peireira et al. 2002; Kaika 2003; Alexandre 2005; François 2006; Margat 2006; Pearce 2006; Bucknall 2007; Barnes 2009; Garcier 2010; Lavie et al. 2015). Here, the authors demonstrate that the management of water is not just an environmental problem, but also is largely influenced by climatic variations or new actors. Julien Charbonnier, an archaeologist (Chapter "The Genesis of Oases in Southeast Arabia: Rethinking Current Theories and Models"), positions himself in the debate on the age of the oases of the Southeast Arabian peninsula based on the issue of climatic variations since the second half of the Holocene, which enables him to re-assess the current and past practices in irrigation management. From when do these oases date? Have the irrigation techniques evolved as a function of changes in the climate? What types of agrosystems were used? Camille Clément, also an archaeologist (Chapter "The Oasis of the Chicama Valley: Water Management from the Chimú to the Spaniards (Eleventh to Seventeenth Century AD) on the North Coast of Peru"), shows that water management is one of the elements of development or of relative decline of the Peruvian oasis of Chicama. Although the current landscape results from the choice of new systems of production (sugar cane, haciendas) and management (*encomiendas* for example) by the Spanish colonisers in the sixteenth century, the peak of the oasis was due to the development of an irrigation system by the Chimu, a pre-Inca civilisation who occupied the region in the eleventh century. As for David Blanchon (Chapter "Who Runs the Orange River Oasis? A Case Study of the Midstream Orange River Oasis, Northern Cape Province, South Africa"), he takes us to the linear oasis of the Orange River in South Africa, where the opening up to international markets at the end of apartheid in the 1990s led to the production of grapes for export to European customers in winter. However, the national policy to divert water from the main tributary, the Vaal, towards Johannesburg and Pretoria could endanger the new irrigated areas.

Part III *Global Control at Different Temporal Scales* highlights some of the long-term consequences of a sudden upheaval. The approach chosen is the effect of the arrival of new actors. The four case studies presented are used to investigate the weight and impact of new interactions on the transformation of networks of actors and spatial structures. Hermann Kreutzmann and Stefan Schütte (Chapter "The Baghlan Oasis in Transition—from Autocratic Modernisation to Contested Spaces") have studied the links between the German military intervention, the drying up of a swamp to create irrigated areas for sugar production, and a sugar factory in the Baghlan oasis in Afghanistan. This example of the transformation of a swamp into an oasis demonstrates once again the many types of oasis. Emilie Lavie and Monique Fort (Chapter "Mountain-Oases Faced with New Roads: Case Studies from the Andes and the Himalayas") have chosen to observe the recent changes in two oasian areas situated in high mountains close to international borders, the oasis of the town of Uspallata in Argentina and the oasian archipelago of Mustang District in Nepal, via the opening of a road enabling the development of exchanges with the great economic powers of Chile, Brazil and India (and eventually China). Irène Carpentier and Alia Gana (Chapter "Changing Agricultural

Practices in the Oases of Southern Tunisia: Conflict and Competition for Resources in a Post-Revolutionary and Globalisation Context") analyse the new trajectories of the oases of Gabès-Chenini and Tozeur since the Arab spring: in fact, the changes begun before the revolution, such as the development of ecotourism, the adaptation of agriculture to the international market, new methods of water and land management and the valorisation of the heritage of oases, have been re-energised following the change in regime. Philippe Cadène (Chapter "Tourism Development in the Thar Desert, Rajasthan") discusses the changes occurring in a globalisation context using a strictly urban example: the city-oases of the Indian Thar Desert. Here, the issue is not agriculture, but tourism and urban dynamics linked to significant growth since the 1980s.

Lastly, echoing the temporal changes highlighted in the second and third parts, the fourth part, by way of a conclusion, describes three examples of spatial transformations. From a wholly geographical aspect, *Marginalisation and spatial reversal* presents different case studies to reveal the origin of multiscale changes and their roles; at the local scale, in the structures and internal exchanges of the oasis, and at the regional scale, in inter-oasian exchanges. What are the new inter-oasian networks? What relationships are being established with other areas? Thus, according to Frédéric Alexandre (Chapter "Insular Oases in Globalisation: The Ribeiras of the Cape Verde Archipelago, Fragmented and Fragile Areas on the Way to Marginalisation"), the Cap Verde *ribeiras*, island agricultural areas, first suffered from "superinsularity" under the dictatorship of Salazar. However, while some have benefited from the rise in tourism following independence, others are experiencing unemployment or socio-economic problems and remain marginalised. For Alain Cariou (Chapter "Liwa: The Mutation of an Agricultural Oasis into a Strategic Reserve Dedicated to a Secure Water Supply for Abu Dhabi"), the oasis of Liwa in Abu Dhabi, which constituted the heart of exchanges in the region before the oil era, has also lost its political function in favour of the urbanisation of the coast and has been relegated to a strategic water reserve for the cities, becoming peripheral to central decision-making. Lastly, Emilie Lavie, Santa Salatino and José Morábito (Chapter "The Transformation of the Oases of Mendoza (Argentina): How the Provincial Socio-Spatial Structure was Reversed by the Crises of the 1980s and 2000s") use the oases of Mendoza in Argentina to illustrate how a wine market crisis transformed a desert into a centre of economic power, marginalising the ancient oasian areas. Finally, the authors of these three chapters demonstrate how a trigger (tourism, oil and a wine market crisis, respectively) has transformed systems with a long history, reversing central–peripheral structures at both regional and intra-oasian scales.

The map below presents the different case studies in this book (Fig. 1). The objective of this work was not to provide an exhaustive list of oases undergoing

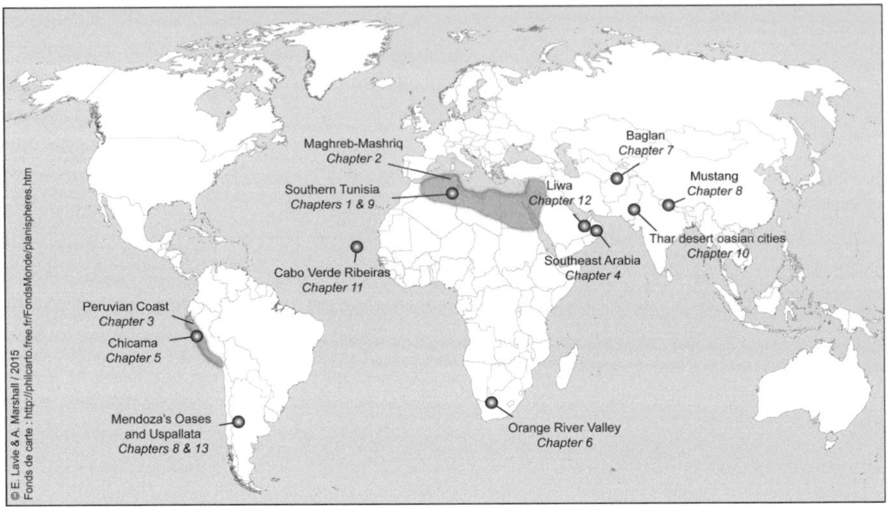

Fig. 1 Location of oases studied in the book (E. Lavie and A. Marshall 2015)

change but to give an updated picture of the diversity of oasian lands using a multidisciplinary approach. Through the lens of globalisation, changes at various spatial and temporal scales have been investigated in order to re-assess oasian areas.

Anaïs Marshall
Emilie Lavie

References

Aguilera-Klink F, Perez-Moriana E et al (2000) The social construction of scarcity. The case of water in Tenerife (Canary Islands). Ecol Econ 34(2):233–245

Alexandre O (2005) Lorsque la pénurie fait ressource. Etude géographique de la notion de ressource. L'exemple de la politique de l'eau au Maroc. Géocarrefour 80(4):263–272

Bakker K (2000) Privatizing water, producing scarcity: the Yorkshire Drought of 1995. Econ Geogr 76(1):4–27

Barnes J (2009) Managing the waters of Bath Country: the politics of water scarcity in Syria. Geopolitics 1–37

Battesti V (2005) Jardins au désert, Évolution des pratiques et savoirs oasiens, Jérid tunisien, Éditions IRD, coll. À travers champs, Paris, 2005, p 440. http://halshs.archives-ouvertes.fr/docs/00/61/48/34/PDF/Jardins_au_desert-Battesti_2005_halshs-00004609v2.pdf

Bucknall J (1997–2007) Making the most of scarcity: accountability for better water management results in the Middle East and North Africa, World Bank, p 235

Clouet Y (1995) Les Oasis. Mappemonde 4(1995):44–48 (en PJ)

Dollfus O (1997) La mondialisation. Paris, Presses de la Fondation nationale des sciences politiques, p 167

François M (2006) La pénurie d'eau en Espagne: un déficit physique ou socio-économique? Géocarrefour 81(1):25–35

Garcier R (2010) Du bon usage de la pénurie en eau. Pollution, pénurie et réponses institutionnelles en lorraine, 1949–1971. Géocarrefour 85(2):169–180

Garcier R, Bravard J-P (2013) La durabilité des oasis: aperçus de géographie historique à partir du cas de l'oasis de Kharga, Egypte. Colloque "Oasis dans la mondialisation: ruptures et continuités", Colloquium "Oases in globalisation: ruptures and continuités", Colloquio "Los oasis en la globalizacion: rompimientos y continuidades", Dec 2013, Paris, France. pp 29–36, Actes du colloque "Oasis dans la mondialisation: ruptures et continuités"

Garcier R, Bravard J-P (2014) Qu'est-ce qu'une oasis? Réflexions géographiques sur un objet-limite. In: Tallet et Zivie-Coche (dirs.) Le Myrte et la Rose—Mélanges offerts à Françoise Dunand, Presses Universitaires de Montpellier, pp 305–323

Gervais-Lambony P, Landy F (2007) Introduction. Autrepart 1(41):3–14. www.cairn.info/revue-autrepart-2007-1-page-3.html

Kaika M (2003) Constructing scarcity and sensationalising water politics: 170 days that shook Athens. Antipode 35(5):919–954

Lacoste Y (1990) Oasis dans Encyclopædia Universalis, vol XVI. Paris, 1990

Lavie E, Maupin A, Blanchon D (2015) La pénurie de l'eau: un discours à quelles fins? In: Redon M, Magrin G, Chauvin E, Perrier-Bruslé L, Lavie E (Dir.) Ressources mondialisées, essai de géographie politique. Paris, Publications de la Sorbonne. p 336. pp 143–166

Lombard J, Mesclier E, Velut S (éd.) (2006) La mondialisation côté Sud. Acteurs et territoires. Paris, IRD-Éditions rue d'Ulm, p 496

Margat J (2006) Quels indicateurs pertinents de la pénurie d'eau? Géocarrefour 80(4):261–262

Mainguet M (2003) L'homme et la sécheresse. Paris, Masson géographie. p 335

Metha L (2001) The manufacture of Popular Perceptions of Scarcity: dams and Water related Narratives in Gujarat, India. World Development 29(12):2025–2041

Pearce F (2006) When the rivers run dry: water, the defining crisis of the twenty-first century. Beacon Press. p 224

Pereira LS, Cordery I, Iacovides I (2002) Coping with water scarcity. IHP-VI Technical Documents in Hydrology No. 58. Paris, International hydrological programm, UNESCO, p 269

Postel S (1992) The last oasis: facing water scarcity. Earthscan Publications, Londres. p 219

Troin J-F (2005) Iles et oasis: de l'isolat au monde. Annales de Géographie 114(644):339–341

Part I
Oasis?

Oases in Southern Tunisia: The End or the Renewal of a Clever Human Invention?

Bénédicte Veyrac-Ben Ahmed and Slaheddine Abdedayem

Abstract Today, we do not know how to define oases or how to classify them due to the diversity of their situations and changing routes on which they are situated. We may eventually keep this name for the small islands of greenery now irrigated by wells. They may have lost their trees, and may have been cut off from their network of exchange, or even been abandoned by farmers who have become urban dwellers. Do the new features of these spaces completely transform the age-old notion of oasis, even if the oasis landscape remains? Is then the notion of an oasis only linked to its superficial image? In looking for a more modern and dynamic definition of these spaces, it is relevant to propose new typologies. Once we assume that future oases will never look like they did in the past, this may imply not just an evolution of the term, but its demise.

Keywords Oasis · Tunisia · Evolution · Traditional landscape

In the last twenty years, some oases have become the focus of programmes of safeguard, protection, acknowledgement and even enhancement. These local and international projects highlight the cultural and social characteristics of such spaces as well as their great biological and genetic diversity. In fact, they are isolated in a hostile environment but, at the same time, part of a very extensive network of exchange. Oases are also the site of specific knowledge developed by native peoples. However, these sources of water have undergone significant changes since the second half of the twentieth century, changes that began earlier but had remained less visible. Can we speak of the degradation, decline or even death of oases as

B. Veyrac-Ben Ahmed (✉)
Université de Toulouse, UMR GEODE, Toulouse, France
e-mail: b.veyrac-benahmed@orange.fr

B. Veyrac-Ben Ahmed
Université J. Jaurès, MSH, 5 Allées A. Machado, 31058 Cedex 1 Toulouse, France

S. Abdedayem
Laboratoire d'économie et sociétés rurales (LESOR), Medenine, Tunisia

E. Lavie and A. Marshall (eds.), *Oases and Globalisation*,
Springer Geography, DOI 10.1007/978-3-319-50749-1_1

Mainguet does (2003)? Moreover, oases are no longer the only inhabited and cultivated areas in the immensity of the desert.

Thus, as the oasis landscape in southern Tunisia has greatly evolved, and its purpose seems to have changed a great deal, with oases no longer being the only "daughters in the desert", we may wonder whether the term oasis is still relevant. Are oasis spaces about to disappear or are they already defunct? Or are they a concept that needs updating? In the southern Tunisian local language, as well as in official documents, different terms are used to refer to "traditional oases", "modern oases" or "new irrigated areas". At the same time, does the definition of an oasis correspond to what it was before and not what it is today? A more modern and dynamic definition seems essential, in order to take into account the various types of landscape. Such a more open-minded outlook could help provide a more specific and efficient management of these areas at every level. Is the term oasis still adapted to the varied states of the landscapes? Are we witnessing the death of a clever human invention or are we on the brink of a rebirth but with new shapes and functions?

We shall attempt to answer these questions on the basis of two key elements: first, the age-old definition of an oasis, its use and its image and second, the observation and systemic analysis of oasis areas, their evolution and their functions, mainly in southern Tunisia.

1 The Use of an Image Suited to a Consensual Definition

An oasis conjures up visions of travelling, a change of scenery, associated with recurrent elements: palm trees, the presence of water and possibly the presence of men through music and traditional artefacts. Spontaneously, the layman sees an oasis as an untouched area, a potential refuge for men in a hostile environment, a place of serenity and eternity and a place where nothing ever changes. It is this mixture, this ambiguity, which is put forward when the image of an oasis is used for tourism and trade purposes (dates). Posters in tourist offices and travel agencies show photographs of oases to promote genuine natural locations in the middle of the Sahara desert or on its fringe.

1.1 A Basic Definition of an Oasis: As Always it is a Threefold Water–Men–Vegetation System

Oases have been described and analysed from an agronomic as well as from a social point of view. Their biological wealth (great diversity of vegetation and animal life, endemic species) has only been acknowledged internationally more recently. Many writers have shown interest in oases, either to understand them as a whole or to describe and analyse some of them in particular.

During the seminar on the agricultural system of the Nefta oasis in 1988, several typological elements were established as being particularly important to define oases and their agricultural systems: implantation in an arid environment, the existence of a microclimate linked to a man-made cultivation in strata, an intensive but isolated agro-ecosystem and a place where a settled population lives and where economic and social cultural activities are intense in a desert environment (according to Skouri 1990). To complete such elements, Kassah (1996, 2010) put forward the diversity of oases in the world. He underlined the permanence of their situation in arid or semi-arid environments, the fact that they are threefold water–men–vegetation systems, and their multiple functions. However, at the local level, the borders between oases and desert landscapes are sometimes hard to define. As time has gone by, elements have been added and others have disappeared. Battesti (2005) emphasised the difficulty of marking the fluctuating boundaries of such locations.

We focus on the threefold idea. First, it enables an oasis to be differentiated from the spontaneous presence of vegetation in a desert landscape. This man-made landscape is a complex and specific anthropo-ecosystem. Although the palm tree is the first visible landmark, it is associated with other vegetation in three vertical layers. The highest stratum is constituted by palm trees. It plays an essential part in the protection of cultivated lands by creating a wetter microclimate than in the outer boundaries, limiting solar radiation and the drying effects of the wind. Besides, irrigation maintains a certain level of moisture. The second intermediate stratum is composed of other species of fruit tree such as lemon, orange, pomegranate, peach, apricot and olive trees. The last and lowest stratum consists of market gardening (carrots, lettuce, onions, etc.), fodder such as alfalfa and industrial crops (tobacco and henna).

Secondly, in the places where rainfall is scarce and where surface water is rare, irrigation is necessary to grow crops. The maximum average rainfall in southern Tunisia occurs in Gabès with 180 mm a year. Such irrigation is made possible by the collective management of water resources. It is used to come from springs, natural resurgences of underground water rising to the surface or *foggara*, a system of draining galleries in mountains. Water is then driven through the oasis by a system of *seguias* or irrigation channels.

As a result, the very shape of an oasis is defined by its water system and distribution network. Moreover, the ownership of the best plots of land reveals the power of the rich local families. Their plots are situated next to the main dispatcher leading to the major *seguias* giving easy access to the water resource. Thus, the shape of an oasis can be considered to reflect the social organisation of the community (owners managing the water).

Consequently, man is at the root of oasis landscapes. Oases are defined by the tangible and marked presence of vegetation under control and a complex system of irrigation and drainage.

1.2 A Specific Anthropo-Ecosystem

An oasis is a common space for multiple uses and functions, where recurrent elements are water/man/vegetation. They structure the landscape with the aim of

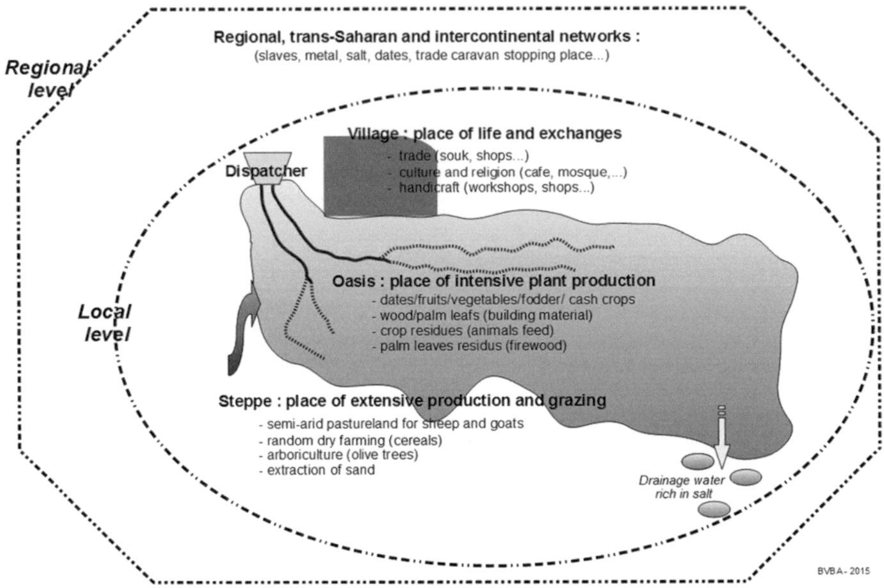

Fig. 1 Spatial positioning and organisation of traditional oasis system at the end of nineteenth century

production. An oasis is thus a place to produce crops and also rear animals on the fodder grown for the flocks that graze outside the oasis limits. An oasis is also a location of handicraft industries (wickerwork, pottery and woodwork).

Thus, complementary activities illustrate the connection between human settlements (the village), the nearby environment (oasis) and the overall environment (steppe) (Fig. 1). Such a diversity of production makes it a favoured place of exchange with the outside world. Due to all these factors, oases fit in with an economic, social and cultural network of commercial exchange, both trans-Saharan and Mediterranean. Even if oases are defined in relation to a much wider space, they have some specific characteristics, a particular farming population and a typical social organisation.

An oasis is a very special "anthropo-agro-ecological" system where man plays a central part, shaping the landscape horizontally (complementarity of steppe/oasis, urban spaces and rural spaces) and vertically (3 vegetation strata). To sum up, not only is man a factor in the evolution of oases but also an essential element in their creation, giving these ecosystems their originality. Without man, there would be no oases.

Because oases have been relatively isolated for so long, they have developed their own characteristics: there is great genetic biodiversity among species (i.e. a variety of palm trees on a regional scale). In many oases, some varieties are endemic. However, this diversity also has a strong anthropic character. In an environment where natural conditions are particularly hard, even hostile, men have been forced to developed ingenious systems of adaptation. Thanks to intercontinental and transcontinental networks of exchange, men have discovered and imported species and techniques that they have gradually adapted to meet their requirements. Such adaptations have taken centuries, as in the case of alfalfa. “Gabesi” alfalfa can now grow despite being irrigated by very salty water (2 g/l).

1.3 An Anthropo-Ecosystem Identified as an Inheritance?

Oases appear to be unchanging but in fact they are not. This apparent stability is linked to the constant search for a balance between the environment and man, and between resources and constraints. The recent and very rapid changes seem to have upset and destroyed this apparent stability.

Today, oases no longer correspond to the definition given in the 1980s, which does not take into account recent developments. There is a discrepancy between the present-day state of oases and what international organisations want to preserve as an inheritance of the past. They wish to safeguard the genetic and cultural heritage of oases, but they refuse to see that their degradation is well underway.

So, are oases still a reality or are they definitely defunct? Must we limit the use of the term of oasis to the threefold system of water–man–vegetation, and do away with the consequences of interactions between these elements? If we retain the term oasis only in relation to the threefold system, then all the irrigated agricultural areas, including the more recent ones, can be named oases, at least those in the making. With their recent changes, can the oldest oases, such as “traditional oases” in southern Tunisia, still be defined as oases?

2 Recent Expansions and Age-Old Creations of Irrigated Areas in Southern Tunisia

In southern Tunisia, in the last century and particularly in the last few decades, irrigated agricultural areas have greatly expanded: from 49,421 acres (nearly 20,000 hectares) in 1962 to 98,842 acres (nearly 40,000 ha) in 2010 (Kassah 2010). In the region of Gabès, irrigated areas increased from 6671 acres (nearly 2700 ha) to 33,853 acres (nearly 13,700 ha) between 1900 and 2005 (Abdedayem 2009). Agricultural land has become multiform and complex.

2.1 Diversity of Irrigated Agricultural Areas

Deep drilling techniques have enabled cultivated areas to be established due to irrigation techniques in the steppes. Formerly, these zones were only used for extensive breeding or even for dry culture (very haphazard).

The term oasis is thus mainly used to describe those that already existed at the time of colonisation and are called *oasis traditionnelles* (traditional oases). The new forms of exploiting irrigated plots that are developing in southern Tunisia are in turn called *oasis modernes* (modern oases) and then *nouveaux périmètres irrigués* or NPI (new irrigated areas).

Irrigated lands are increasing, and farming techniques for exploiting steppe zones are diversifying, and yet the new ways of working the land have not replaced earlier ones. They coexist side by side with them or create new small islands of cultivation. Based on the dates of irrigated areas appearing in official documents, they can be classified into three distinct groups. Several expressions in everyday language enable these areas to be differentiated.

2.1.1 Traditional Oases

Traditional oases were created before 1900. These areas represented 6671 acres (nearly 2700 ha) in the region of Gabès and 16,062 acres (nearly 6500 ha) in the whole of southern Tunisia. It is about the same nowadays. Their water supply used to come mainly from springs. The supply was managed collectively by an association of landowners and then by farmers. Most of these springs have run dry since 1980 or 1990. In this type of area, a landscape with three vegetal strata is usually found. The surface is divided into small plots like a honeycomb because of successive inheritances and is clearly visible on aerial photographs. The Gabès coastal oasis, Chebika and El Ksar mountain oases and the Tozeur Saharan oasis are found in this classification.

2.1.2 Modern Oases

These were created after 1900, due to new deep drilling techniques by colonial authorities and then by the Tunisian state. Water is managed collectively by a farmer association subsidised by the state. It used to be called *Jamia el mâa*, then AIC, GIC and now GDA (Organisation for Agricultural Development). The area was managed rationally to ensure a better output: the fields were laid out like a chessboard. Palm trees were aligned and paths drawn in a straight line to enable easy access to plots and the water supply. These modern oases represent 10,625 acres in the Gabès region and 61,825 acres in southern Tunisia. The spreading zones of Tozeur and Nefta, where the cultivation of the Deglet Nour is practically a single crop, are part of this category. The Deglet Nour is a variety of date that is particularly appreciated for export since it brings in foreign currency.

2.1.3 New Irrigated Areas

After 1970, some new irrigated public and private areas were created around surface wells, individual or collective drilling or alternative water supplies. Water supplies are often managed privately, but farmers can combine the use of both individual and public systems. For the Gabès region, these new irrigated public areas represent 7487 acres (nearly 3030 ha), while new irrigated private ones are estimated at 9068 acres (nearly 3670 ha). To this group can be added the recent spreading zone of Ben Ghilouf in El Hamma (geothermic greenhouses) and private irrigated areas created by Ghannouchiens (Ghannouch native farmers) around 1990, like in Tenesli created at the beginning of 2000.

2.2 Spatial Organisation of the Various Irrigated Perimeters

The area occupied by earlier oases is currently smaller than other irrigated areas. Modern oases have provided access to ownership for new groups such as former nomadic populations and farm workers, also called *khammès*. For this reason, they are often found in the vicinity of earlier oases. The new irrigated areas are either next to earlier islands of greenery or perimeters created *ex nihilo* (Chenchou, Kettana).

According to the definition of oases given previously, three strata of vegetation are essential to the genuine landscape. Everyday language classifies oases by their creation date. In that case, there would be a match between the term "traditional oasis" and that of oases as defined previously. However, traditional oases have greatly evolved through the ages, as shown by the strata and diversity of vegetation, the composition of the soil and traces of social organisation of the water supply. In fact, by their creation date and their characteristics, they correspond to what is expected of an oasis.

However, in southern Tunisia, if the geographical distribution of irrigated areas is compared according to their status or their age, or even their landscape characteristics, they should be similar, but this is not always the case. To understand the reason for such differences, it is necessary to describe and analyse their diverse changes. Incompatibility between age and landscape could question the supposed linearity in the evolution of oases.

3 The Oldest Traditional Oases and More Recent Ones: Towards a Simplification of the Multistratum System?

Traditional oases are often composed of an earlier historical nucleus, which can be referred to as the "mother oasis" as opposed to extensions called "daughter oases" (Abdedayem 2009). These later ones can be related to modern oases as they have been irrigated since they were set up thanks to deep drilling. Generally, they are no more than a century old (created during the colonial occupation). Such extensions may have expanded gradually over the course of time, but they were always set up

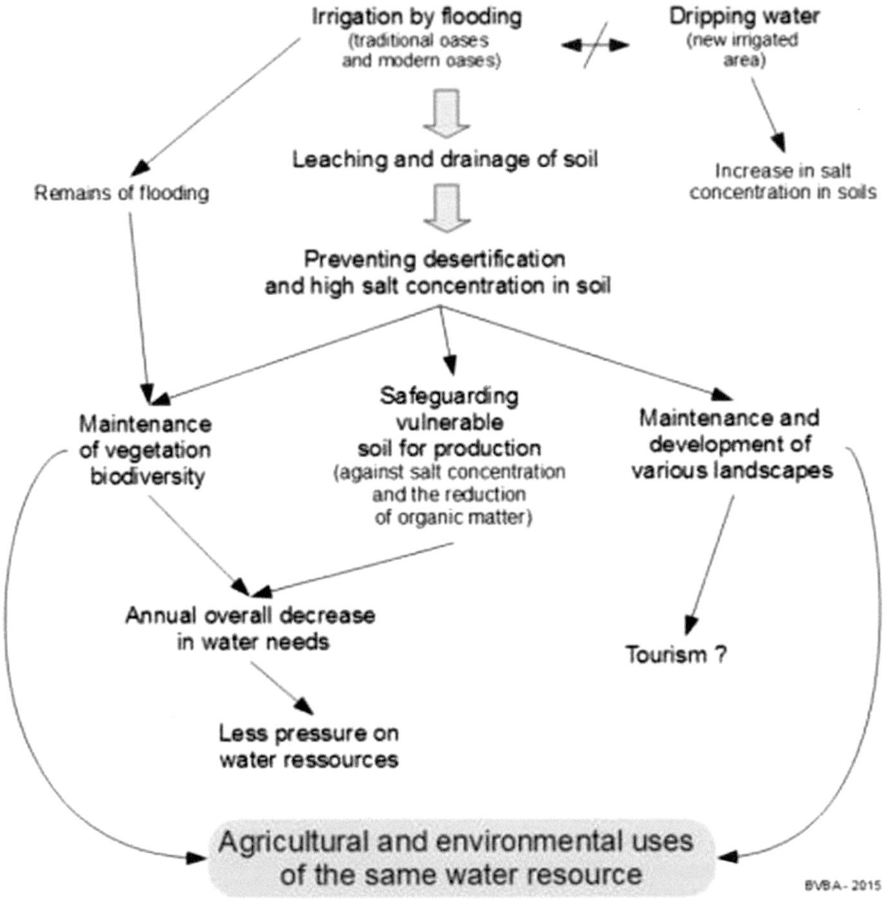

Fig. 2 Agricultural and environmental enhancement of the water resource in traditional and modern oases via traditional techniques of irrigation (flooding)

in the immediate vicinity of neighbouring plots of land. The older such extensions are, the harder it is to tell what is historical in them since their landscape is so similar. The division of the plots of land makes access difficult for some *fellahs* (farm workers) in modern oases as well as in older ones. Exploitation and agricultural techniques mainly associated with irrigation are similar (Fig. 2).

Only more recent extensions are visible to the naked eye, due to their smaller number of palm trees and the use of modern agricultural techniques (mechanisation and drip watering). However, they can be confused with earlier renovated plots, reorganised in order to increase production. The deliberate reduction in palm trees lets the light penetrate the plantation. It minimises the competition with other vegetation strata with a view to providing a more profitable crop.

In fact, the cultivation system of an oasis is not becoming more intensive since it is already intensive by nature. The issue is rather a complete change in the objectives of production. It is now aimed at the regional market or urban customers whose needs have become standardised. For instance, they would rather buy a fashionable brand than cheaper local seeds. Farmers are becoming increasingly dependent on hybrid seeds. It is easier to find carrot seeds from "Vilmorin" than "Arbi Carrots", local seeds that are more resistant and require less watering.

The three-stratum landscape in traditional oases is thus less and less obvious owing to the evolution of cultivation techniques and farmers' own decisions to choose other options. Previously, slow evolution and isolation had enabled a great variety of crops as well as endemic species to develop. However, the decline in the consumption of varieties of dates too basic or unfit to be exported has resulted in the removal of many palm trees, mainly in coastal oases. In fact, ideal conditions of maturation for the Deglet Nour variety are not found in such areas. Palm tree diversity specific to coastal, mountain or Saharan oases has tended to shrink considerably. Such a phenomenon is also true of varieties of trees and gardening crops, which are threatened by extinction (Ben Salah 2012).

In some cases, earlier or traditional oases have turned into a two-stratum system: palm trees with fruit trees or fodder plants, or fruit trees with fodder plants as in Mareth. There, palm trees are gradually being replaced by pomegranate and olive trees.

However, most oases have been assimilated into irrigated areas and managed according to the "concept of PPI" (PPI: irrigated public areas) owing to the drying up of natural springs and their replacement by deep drilling. As a result, they have started to lose their special features and other functions they used to fulfil throughout their history. Traditional oases were considered archaic, non-profit-making and wasteful for natural resources (Kassah 2002). The concept and principle of the self-management of water resources by the community is used to comply with the inalienable rights accepted and respected by all. It enabled everyone to have access to free water. This form of management has been replaced by new rules dictated by local government and national government. The focus is on intensive agriculture, economic growth and speculation (monoculture, speculation on land sales), which are detrimental to other aspects specific to oasis ecosystems. Thus, the rules and principles of the setting up and management of natural resources have been disrupted (social organisation, management and cultivation techniques).

Small farmers, since they do not have access to large financial investments, cannot compete with a commercial market from which they are excluded. As an indirect result, they cannot maintain a viable activity. Consequently, multifunctional agriculture in oases is disappearing (shaping of the landscape and social and cultural environment).

This evolution of older traditional oases, considered genuine, has resulted in their losing some of their specific landscape identity. Moreover, these areas are now in competition with other activities, mainly residential or industrial. The proximity of urban centres increases the demand for building land, as well as more competition for water consumption and pollution of the environment, harmful to the oasis.

4 New Irrigated Areas: An Easier Implantation System for Easier Production?

The most modern oases, dating back to the post-colonial period, have often been set up in the vicinity of traditional oases, without a spatial continuity between the two stretches of land. New irrigated perimeters started to be created in the 1990s, mostly on semi-arid pastureland, some far away from old oasis nuclei. They were established following hydro-geological and oil exploratory drillings by the government or by private investors with or without prior legal authorisation.

These new irrigated perimeters are very different from traditional oases and older irrigated perimeters in their management, size, existing production and also farmer investment and involvement. Production in these areas is mainly single stratum. It is intensive, mechanised and uses a great deal of chemicals and more or less developed technologies (e.g. irrigation by drip watering and geothermic greenhouses in order to produce out-of-season tomatoes in Chenchou). In fact, the land is exploited like a mine, that is to say until available natural resources (water and soil) are exhausted.

The targets of such production are the regional, national or even international markets. Cultivated species and varieties are limited to those whose sale is more profitable: potatoes, tomatoes, peppers and onions. Varieties are often hybrids. The aim is to produce graded homogeneous crops. As a result, genetic diversity is very low. It can sometimes be increased by introducing alfalfa to enrich the soil or mitigate the adverse effect of salt in order to enhance the production of plots too rich in salt. Owners can request farm workers to plant fruit trees or olive trees when setting up a market gardening enterprise.

However, in these perimeters, the most favoured vegetation stratum is expected to generate crops corresponding to the standards required by the national and international agro-alimentary market without delay. The second stratum, when it is there, is rather neglected. It is only tolerated on condition that it does not compete with the commercial stratum. This second stratum can be used to supply food for the family. They benefit from the irrigation and the farm work required by profit-making crops. Thus, the areas of Zerkine 1 and 2 usually associate fruit growing with vegetable growing or the production of fodder (tomatoes, onions, garlic and alfalfa).

Finally, the initial one-stratum system is made more complex by the exploitation of a second stratum. It could be considered an alteration of the initial agronomic planning and so a failure of the laying out of the land. However, some farm workers enjoy a more personal involvement, which goes beyond just a financial investment. So, within new irrigated perimeters, various situations can be found side by side:

some plots are partly or totally neglected, whereas others are well cared for. This adaptation of the one-stratum system makes one wonder whether these perimeters could potentially evolve towards a more complex agro-ecosystem and a management similar to the older oasis system.

Nevertheless, one of the main characteristics of these new perimeters remains their low initial biological diversity based on the supply of outside cultivated species, often hybrids, and their high consumption of non-renewable and overexploited resources (water and soil). The evolution towards a multistratum system increases the number of species cultivated without favouring diversity among species.

5 Irrigated Areas Versus Oases?

In the context of rapid evolution and profound changes, irrigated areas in southern Tunisia show a great diversity of shapes and functions. The recent changes in the various perimeters blur the identification of the landscape. Those undergone by traditional oases question the very definition of an oasis and the way it is run. Indeed, traditional oases bear a both a social and a biological legacy. Thus, they can be considered real anthropo-ecological resources through their genetic diversity and the knowledge and management of this diversity by the population concerned.

Whether the evolution of oases is considered an adaptation or a direct disappearance, it is associated with several constant elements. The oasis agro-ecosystem loses its complexity and is weakened when faced with the constraints of an adverse environment and the impact of climate changes to come. The local genetic heritage, with its diversity and its ability to adapt developed over the course of time, represents a very important genetic potential that can be coveted by some people. Although the argument for the urgency of safeguarding a rich ecological legacy has been put forward, the tendency to recording oases as an international heritage corresponds to a social requirement, which takes into account a new use of the landscape for the tourist trade. However, this so-called conservation of the heritage ignores the causes and the continuation of oasis degradation. An "anthropo-agro-ecosystem" is being set aside and safeguarded at all costs without taking into account relevant outside elements linked to this system.

The decision to set up new irrigated perimeters based on rational agronomic planning is a factor of evolution. However, interweaving these spaces (steppe, traditional oases, modern oases and new irrigated perimeters) and the existing relationships between the functions attributed to them are very important. This is one of the keys to understanding overall and regional evolution and that of oases in particular. In fact, all the roles played by oases, until the end of the colonial era, are now shared between various perimeters, different locations and entities, which are in competition (Fig. 3). This competition is linked with the use of a single water resource: the Djeffara (regional complex of middle depth water layer). The functions of an oasis are used to be complementary and set in one local area. Nowadays,

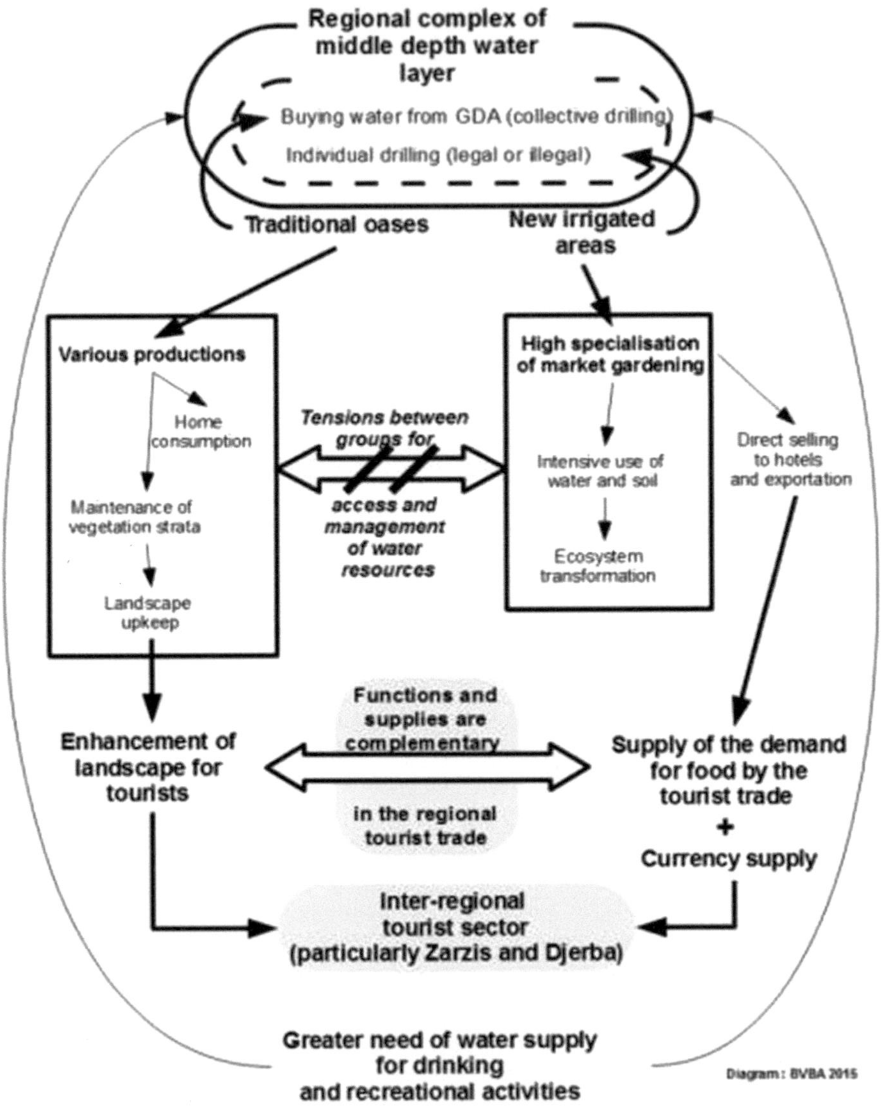

Fig. 3 Division of functions of agricultural areas and competition around the use of a common resource in the region of Gabès

they are spread out over a regional scale. As a result, the complexity of the oasis system is impoverished. Oases could become agricultural areas among others.

By losing their multiple functions, oases have shed a key element of their identity, which is used to define them as unique entities. Consequently, can we still use the term oasis in southern Tunisia? Should they not rather be called "areas with some oasis traits"?

References

Abdedayem S (2009) Mutations socio-spatiales et modes de gouvernance de l'eau dans les oasis périurbaines du gouvernorat de Gabès (Sud-est tunisien), thèse de doctorat de géographie, Université Paris 10, 440 p

Abichou H, Jouve A-M, Rey Valette H, Sghaier M (2009) La valorisation du patrimoine et nouvelles alternatives pour un développement local durable dans le Sud-est tunisien, New Medit n°4/2009, pp 55–62

Aspe C (2012) De l'eau agricole à l'eau environnementale. Résistance et adaptation aux nouveaux enjeux de partage de l'eau en Méditerranée, collection Update Sciences et Technologies, édition Quae, 380 p

Aspe C, Jacqué M (2012) Environnement et société. Une analyse sociologique de la question environnementale, collection Natures sociales, éditions de la Mison des sciences de l'homme, Paris et éditions Quae, Versailles, 279 p

Battesti V (2005) Jardins au désert, évolution des pratiques et savoirs oasiens dans le Jérid tunisien, collection A travers champs, éditions IRD, Paris, 440 p

Battesti V (2009) Tourisme d'oasis. Les mirages naturels et culturels d'une rencontre? Cahiers d'études africaines, 2009/1, n°193–194, p 551–582

Bechraoui A (1980) La vie rurale dans les oasis de Gabès, thèse de doctorat de géographie, Université de lettres de Tunis, 301 p

Ben Salah M (2012) Rapport d'expertise technique sur la biodiversité oasienne en Tunisie, RADDO – ASOC, 76 p

Bonny S (2011) L'agriculture écologiquement intensive: nature et défis. Cahiers Agricultures 20:451–62

Boulay S (2009) Culture nomade versus culture savante. Naissance et vicissitudes d'un tourisme de désert en Adrar mauritanien, Cahiers d'études africaines, 2009/1 n°193–194, p 95–122

Brochier-Puig J (2004) Société locale et Etat face aux limites de la ressource eau (Nefzaoua, Sud-Ouest tunisien), in Environnement et sociétés rurales en mutation: approaches alternatives, pp 307–320

Chanvallon S, Héas S (2011) L'Homme et le Nature: en quête/enquête sensible, Natures Sciences Sociétés, 2011/4, vol 19, pp 355–364

Cheylan J-P (1990) Les oasis sahariennes à foggara: mutations sociales sous fortes contraintes écologiques, in MappeMonde, n°90/4, pp 44–48

Clément M, Meunié A (2010) Inégalités, développement et qualité de l'environnement: mécanismes et application empirique, in Mondes en développement, 2010/3, n°151, pp 67–82

Clouet Y, Dollé V (1998) Aridité, oasis et petite production, exigences hydrauliques et fragilité sociale: une approche par analyse spatiale et socio-économique, in Sécheresse, 2(9):83–94

Colleyn J-P, Devillez F (2009) Le tourisme et les images exotiques, in Cahiers d'études africaines, 2009/1, n°193–194, pp 583–594

Cormier-Salem M-C, Roussel B (2002) Mettre en patrimoine la nature tropicale: une histoire ancienne, des enjeux nouveaux. In: Cormier-Salem et al. (coord) Patrimonialiser la nature tropicale. Dynamiques locales, enjeux internationaux, collection Colloques et séminaires, Paris, pp 15–27

El Fekih P (1966) Les sols des oasis anciennes du Sud tunisien, actes de la conférence sur les sols méditerranéens, Madrid 12–17 septembre, 12 p

El-Belkri AO (1913) Description de l'Afrique septentrionale, in Guckin de Slane, Description géographique du monde connu, Alger, Paris, 405 p

Guérin V (1927) Voyage archéologique dans la régence de Tunis, Paris, Plon, 395 p

Guibert C, Paul A (2013) Usages et intérêts différenciés des espaces maritimes et fluviaux patrimonialisés. Les cas de Saint-Gilles-Croix-de-Vie en Vendée et du canal de Nantes à Brest, in Norois, 2013/3 (n°228), pp 89–101

Guillerme S (2003) L'agroforesterie paysanne kéralaise. Un patrimoine naturel et culturel menacé par le développement de l'hévéaculture. In: Cosaert P et Bart F, Patrimoines et développement dans les pays tropicaux, collection Espaces tropicaux n°18, DYMSET, pp 269–277

Hosni E (2000) Stratégie pour un développement durable du tourisme au Sahara, Décennie mondiale du développement culturel, UNESCO, 68 p

Jeanneaux P, Dare W (2012) Conflits d'usage et processus de patrimonialisation des espaces résidentiels. La Réunion et le Puy-de-Döme, in Economie rurale 2012/6 (n°332), pp 47–59

Jouve A-M (2006) Les trois temps de l'eau au Maroc. L'eau du ciel, l'eau d'Etat, l'eau privée, Confluences Méditerranée, 2006/3, n°58, pp 51–61

Kassah A (1996) Les oasis tunisiennes, aménagements hydro-agricole et développement en zone aride, Série géographique n°13, CERES, 346 p

Kassah A (2002) Irrigation et développement agricole dans le Sud tunisien, In Méditerranée, tome 99, 3-4-2002, Le Sahara, cette « autre Méditerranée » (Fernand Braudel), pp 21–26

Kassah A (2010) Oasis et aménagement en zones arides. Enjeux, défis et stratégies. S. Marlet, I. Mekki. Gestion des ressources naturelles et développement durable des systèmes oasiens du Nefzaoua, Feb 2009, Douz, Tunisia. Cirad, 6 p, 2010. <cirad-00496143>

Mainguet M (2003) Les espaces oasiens: oscillations entre décadence et nouveau développement urbain. In: Mainguet M, Les pays secs; environnement et développement, collection Carrefour, éditions Ellipses, Paris, pp 111–143

Marzougui M (1962) Guebes, jannet eddounya (Gabès, paradis terrestre), 307 p. (en arabe)

Picouet M et al (2004) Environnement et sociétés rurales en mutation. Approches alternatives, collection Latitudes 23, IRD éditions, Paris, 410 p

Senil N et al (2014) Le patrimoine au secours des agricultures familiales? Eclairages méditerranéens, Revue Tiers Monde 2014/4 (n°4), pp 137–158

Sghaier M (2010) Etude de la gouvernance des ressources naturelles dans les oasis. Cas des oasis en Tunisie, Union Internationale pour la Conservation de la Nature, UKAid, 69 p

Skouri M (1990) Eléments de synthèse et conclusions. In Dollé V et Toutain G (éd.) Les systèmes agricoles oasiens, Options Méditerranéennes: Série A., Séminaires Méditerranéens; n°11, pp 331–335

Smouts M-C (2005) Du patrimoine commun de l'humanité aux biens publics globaux, in Cormier-Salem M-Ch (coord.), Patrimoines naturels au Sud. Territoires, identités et stratégies locales, collection Colloques et séminaires, IRD éditions, pp 53–70

UNESCO (2003) Le Sahara des cultures et des peuples, Vers une stratégie pour un développement durable du tourisme au Sahara dans une perspective de lutte contre la pauvreté, Paris, 84 p

Veyrac B (2006) Impacts environnementaux de la mise en valeur agricole d'une zone de parcours. Cas de Tenesli Gabès, Sud tunisien. Etude méthodologique. Mémoire de master 2 recherche « Environnement et paysage », IRA Gabès, UTM, 85 p

Veyrac-Ben Ahmed B, Abdedayem S (2011) Incidences de la « modernisation » du réseau d'irrigation sur l'écosystème oasien: le cas de l'oasis de Gabès (Sud-est tunisien), contribution à l'atelier 1 du colloque Usages écologiques, économiques et sociaux de l'eau agricole en Méditerranées: quels enjeux pour quels services?, Université de Provence. http://www.lped.org/actes-du-colloque-eau-agricole/pdf/th1/Com_Veyrac_Ben_Ahmed.pdf

A Travel Through Oases in French and Arabic Literature

Marc Kober

Abstract The Arabic or French novels of Gamal Ghitani, Ibrahim Al-Koni, Pierre Loti, Charles Wallut, Eugène Fromentin, and André Gide are legendary stories located outside of time and linked to a mystical quest. The travel to reach an oasis is symbolical, and the life within it an *oasisnade*, or oasis narrative. As a dream or a mirage, the oasis is an important imaginary representation for orientalist literature linked with the French colonial process. The main problem for the writer is how to describe poetically such a geographical reality as an oasis inside the complementary reality of the desert, using colours, the "écriture artiste", or patterns such as the foundation story, the Garden of Eden, or the maze. The identity of the traveller is completely changed through the encounter with the oasis in a spiritual or cultural way.

Keywords Oasis · Novel · Travel · Quest · Desert · French · Arabic · Writer

1 Introduction

This chapter will focus on four French novels and two contemporary Arabic novels, by Gamal Ghitani (*L'Appel du couchant*)[1] and Ibrahim Al-Koni (*L'Oasis cachée*).[2] The historical context of these novels, apart from their cultural and linguistic differences, has completely changed, but Gamal Ghitani and Ibrahim Al-Koni have

[1]The Call of the West.
[2]The Hidden Oasis.

M. Kober (✉)
Sorbonne-Paris-Cité, EA PLEIADE, University Paris-13, Villetaneuse, France
e-mail: marc.kober@wanadoo.fr

M. Kober
UFR LLSHS, Université Paris-13, 99 Avenue Jean Baptiste Clément, 93430 Villetaneuse, France

E. Lavie and A. Marshall (eds.), *Oases and Globalisation*,
Springer Geography, DOI 10.1007/978-3-319-50749-1_2

not written historical novels, in a realistic way. They create legendary stories about oases, which cannot be understood with reference to the postcolonial situation in Egypt or Libya (Fig. 1). Their novels are located outside of time, in a sort of eternity. A journey is described, but it cannot be clearly seen on maps or on a calendar. Some differences can be seen with Western writers on the subject of oases. In the case of the Arabic novels, the origins of oases are often legendary, although they are not easily located in space. They exist in contrast to deserts as a divine miracle as well as a fragile human creation. The desert and its oases are linked with a mystical quest in Islamic civilisation, as we will see in this essay. The oasis opens up a holy dimension in space. The traveller is a pilgrim who has heard a "call" that will change his life completely, as is clearly represented in the two Arabic novels. This mystical call is not absent from the French novels we chose to read, especially the one by André Gide, but the Islamic aspects are not relevant. Despite the aesthetic and historical differences, the French novels were chosen because of the central importance of the oasis in the story, just like in the modern Arabic novels described above, while the comparison between the two cultures and two distinct periods of time is somewhat interesting, because it shows how the travelogue is a shared creative pattern. Travel literature is a common field, but the travel is immediately symbolic in the case of the Arabic novels, whereas it has a geographical and conquest meaning in the case of those authors very close to the colonial administration of the French empire. However, even Pierre Loti and Eugène Fromentin move from conquest to quest, whatever this might be: a religious or an artistic achievement. Thus, some aspects of their works are perfectly in tune with the contemporary Arabic novels. The reasons are to be found in the physical nature of the oasis itself, as we will try to show. Basically, the geographical

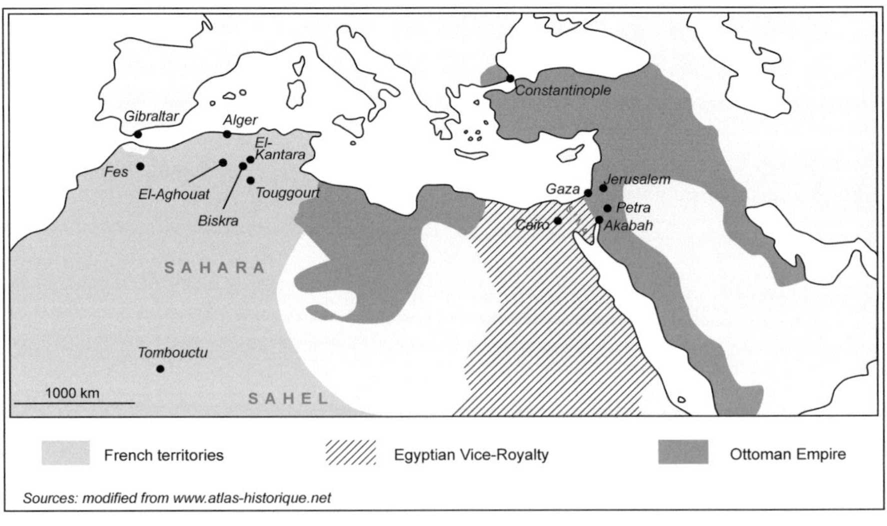

Fig. 1 Maghreb and Mashriq oases in French and Arabic literature. Geopolitical context at the beginning of the 20th Century

characteristics of oases have not changed over time, so they can dictate the same ideas to many writers and produce the same words in order to describe a similar reality.

The French colonial novels about oases, like Charles Wallut's *Grandeur et décadence d'une oasis*, republished as *L'Oasis – scènes du désert* in (1883), are close to the genre of the "Robinsonade", which is a kind of narrative born after the success of Daniel Defoe's novel, Robinson Crusoe, where the action takes place in an exotic context, generally on an island, but, in our view, an oasis is very close to an island in the human imagination. The term first appeared at the beginning of the eighteenth century. The Robinsonade themes are isolation and the origins of civilisation, or human culture. This kind of narrative has a didactic value, especially for educating young people, according to J.J. Rousseau. It is a kind of adventure novel, which uses the main ideas of Defoe's novel, but the Robinsonade also includes the values of colonialism, in the framework of exotic or colonial literature. Our writers are travellers in the Near East, or in Africa close to Europe, like North Africa. They travel to reach an oasis, or they come across oases during their travels. May we suggest another word instead of Robinsonade, such as "oasisnade", in the sense of an oasis narrative? Then, we could have the right word to describe a common literature field. The oasis is an image of colonisation without any metaphysical sense apart from Providence. For Pierre Loti, travelling through deserts between Cairo and Jerusalem, the oases encountered during his travels are linked with a personal image of Eden, purity and sensuality. It is this image that we can find in *Le Désert* (1913). Eugène Fromentin, with *Un Eté dans le Sahara* (1887), gives a very original version of the visual aspects of the oasis due to his artistic status. André Gide, with *L'Immoraliste* (1985), is in search of a moral climate, hot of course, where the human body can be healthy and beautiful at the same time. His hero finds it in Algeria, especially in one oasis, Biskra, after having left southern Italy.

First, we describe the French literary vision, before focusing on a few Arabic examples, in order to see how far their perceptions can be compared, in the common perspective of oasis narratives, or oasisnades.

2 French Colonial Novels

Can we associate the oasis novels with a general reflection about travel accounts in what was used to be called in France "*L'Orient*" (the East), or "Levant", or the Ottoman Empire? In a sense, yes, because oases are linked with the general theme of the desert, the Sahara or Sinai. The Algerian oases and the desert, as a theme, are often described in these travel accounts (Berchet 2001: 1020). The desert, and the oases within it, is a space represented, read, and written about, but it is not very easy to find specific oasis novels, like those written by Bahaa Taher or Ibrahim Al-Koni. This does not mean that they do not exist, but they seem to lack originality, like Charles Wallut's novel, whereas their ideological purpose is obvious. Thus, the desert is a global scene where the oases may appear but as a detail in the picture, not as the main subject. We find a real link between the general rules of travel accounts in the "Orient" and the description of oases as part of the travel. In several famous travel accounts, such as *Un Eté dans le Sahara* or *Le Désert*, the oasis is not the centre of the story, but it plays a very important role, especially for Fromentin, as a pictorial topic, a challenge for both the painter and the writer. Then, in the famous French novel *L'Immoraliste*, the life of the main character finds its deepest truth under the palm trees of Biskra. Through these four French novels, we will see that within a general colonial framework, literary achievements can be very different. The value given to oases can be very personal according to the writer.

2.1 Grandeur et Décadence d'une Oasis *by Wallut (Second Printing 1883)*

This French novel gave birth to a shorter version under the title *L'Oasis – Scènes du désert* (1883). First, we can summarise the novel and read it very quickly; Captain Onésime Lafourche leaves Gibraltar on 1 March 1815, but only reaches Fes in February 1816. The novel thus begins like a travel account in the desert because the captain and his group of fellow travellers cross the desert between Fes and Timbuktu. The Sahara is soon described as a "*mer figée*"[3] (Wallut 1883: 70) which is an Arabic image: the desert as a sea. The author also reminds the reader of the words of Ptolemy, who compares the desert to a panther's fur, yellow-like sand, and black-like oases (Wallut: 71). Generally speaking, orientalist literature is considered a dream, an imaginary representation: the so-called oriental mirage, according to Louis Bertrand (Berchet: 10). Right away, Wallut uses the potential quality of the Sahara: a place where the traveller meets mirages. We can find various statements about hallucinations in chapter VIII: "*Les Phénomènes du désert*". The mirage is a reverse image; for example, the tops of palm trees are seen upside down. The scientific allusion to mirages in the desert precedes the encounter with the oasis announced by the title, as if it was itself a sort of mirage, and a reverse image, which is the main originality of this description. The oasis was abandoned a long time ago and is completely dry. The novel becomes didactic with a typology between three types of oasis. The old oasis used to be irrigated by an artesian well, but this well has been filled with sand. The oasis is dying; it will come to life again because, as quoted in Arabic, "*Bahar toht el erd*" (the sea is under the earth). So, the oasis is symbolically the place where man can be born a second time. The oasis is feminine, like the "Orient" (East) for the whole nineteenth century. It was like a sleeping beauty, and the small group of Europeans will awaken this lady through skilful and technical activity. Their energetic digging (140 m deep!) enables the oasis to come back to life in a few days. Then, the author alternates between a

[3]A frozen sea.

human and a divine vision of reality. The oasis is classically compared to Eden, "*le Paradis terrestre*" (Wallut: 120), and is also named "*La Terre promise*". The Promised Land was reached by the Jews after crossing the Red Sea, just like the small troop left Fes and crossed the Sahara. At the same time, the author describes the cattle grazing and the date palm culture, and the organisation of the small colony with an elected government. So, the oasis is described according to Robinson Crusoe's patterns. It is a Robinsonade, one among many novels written after Defoe's, especially for teenagers. Although the oasis is like a safe island, isolated in the desert of sand (which is another common image associated with oases), several dangers appear: clouds of grasshoppers, ostriches, "*Targui*" or "*Touaregs*", a large desert tribe. This is a typical vision of the colonialist period, when local inhabitants are considered either friends or foes, facing the European power. The haven of peace becomes a battlefield, and the oasis is destroyed. The small group of travellers leaves the oasis after fourteen years. It is now 1830, precisely when Algeria is defeated by French troops. We can deduce from this coincidence that the main argument of this novel, apart from desert exoticism, is to justify, in a mythical way, the French colonial process in Algeria.

2.2 Le Désert *by Loti (1913)*

The trip described by the famous French writer is one of the most interesting among his various travel books, although the general idea is very common too: Chateaubriand invented the "*itinéraire circulaire*", Egypt–Palestine–Lebanon–Asia Minor–Constantinople–Athens–Greece, which was destined to be written about so many times later on (Berchet: 10). The first part of this journey ends at Gaza and Jerusalem on Easter Sunday. Pierre Loti quotes the Bible before every chapter, and he visits the Saint Catherine monastery in the Sinai mountains, but he offers the reader something very different: the beauty of the desert, which is the outermost borders of the world, a typical oriental space (Berchet: 1042). Following the sandy road through Sinai, Akabah, and the Petra desert, the travellers meet the most difficult road to Jerusalem. Once they have climbed Mount Sinai, they shelter for a few days in the "*demeure de la solitude*" (Loti 1913 : 46) or the "*le couvent âgé d'une quinzaine de siècles*" (Loti 1913: 45): Deir Sant Katreen. The famous writer is invited to visit the "*jardin muré*" (69). The gardens of Saint Catherine are built in terraces with high walls around them, near the community's cemetery. Cypresses, olive trees, a vineyard, and citrus trees grow there. Pierre Loti compares his experience with an oriental spring (75), or an artificial garden: "*Et il est singulier, ce printemps-là, qu'on sent venu uniquement pour ce jardin artificiel* […] *puisque nulle part ailleurs, il ne trouvera rien à reverdir, dans l'infini des sables et des pierres mortes* […]" (75). These gardens are not an oasis, but they work as if they were: they are a green, artificial, closed, and isolated area surrounded by dry mountains. Once the travellers have left Mount Sinai, they cross a desert land and come across a few valley oases, such as Oued-el-Aïn, the valley of the fountain.

The main problem is how to describe poetically such a geographical reality as an oasis. However, Loti is a prominent writer because he created for his readers a personal variety of the "*pittoresque exotique*" (Mougin-Haddad-Wolting 2002: 536). In fact, it is not necessary to be really original: the reader accepts with pleasure the same other or "*le même autre*" (Berchet: 10). The repetition gives him/her great pleasure, "*le plaisir de la répétition*" (10). So Pierre Loti uses the same adjectives, "*un lieu enchanté*" (Loti: 90), "*l'oasis charmante*" (93), or the same comparisons with Eden: "*Avec quelles images de fraîcheur empruntées aux poètes de l'ancien Orient peindre cet Eden caché dans les granits du désert*?" (92). This is the beginning of chapter XIX, which repeats the previous chapter, word for word. The oasis exists by its contrast with the surrounding desert. So it is characterised as "small" as opposed to the greatness of the desert: "*En route, le long de la mer, – et sitôt disparue la petite oasis charmante, le grand désert vous ressaisit*" (119). It is a very narrow and restricted area, whereas the desert is described everywhere as "*étendue*": "*Et nous recommençons à faire route vers le Nord, dans ce désert d'un gris jaunâtre qui semble n'être plus rien que l'étendue, – l'étendue sous sa forme la plus simple, mais aussi la plus excitante à courir*" (198). A large stretch or expanse of sand is seen everywhere as the desert area, and within it, the traveller discovers a tiny and closed space completely apart: the oasis. The traveller approaches the complementary reality of the desert and oases according to a physical experience, such as a phenomenology of the desert. Pierre Loti is a modern author in that sense: he tries to find a language suitable for a phenomenological experience. He emphasises the sudden change when one leaves the desert and enters an oasis such as Oued-el-Aïn. The traveller is delighted by this transformation of the landscape, like the opening of the curtain when the play begins: "[…] *quand elle s'ouvre tout à coup comme un décor qui change entre deux hauts portants de montagne*" (90). The world of the theatre and geophysical reality mix to recreate in words the intensity of a unique perception of the changing desert space. The writer may be helped by geology, the volumes and shapes of the rocks, such as walls or circles, and above all by botany. The great pleasure of tired eyes is to gaze at plants and trees, associated with water, because it is a new sensation for the traveller, so pleasant that the author can list the various plants several times. He compares this natural phenomenon with cultivated areas such as gardens ("*jardins sauvages*", or "*bocages*", Loti: 94). Of course, for this special problem of how to describe an oasis, colours are of great help, as we will see with the example of Eugène Fromentin, a painter and writer at the same time. The oasis has a special colour. The rocks are red, but the other components are "painted" using cold colours (blue or green). The oasis is described as "[…] *ce triple cirque de rochers sanglants parmi les verdures bleues*" (Loti: 91). The blue colour is given by the tamarinds, reeds, and palm trees. The restricted area of the oasis is blue, whereas the great extent of the desert is grey and yellow. The latter is even the colour of lizards, "*couleur du sol et de l'étendue*" (198). We can say that this original discourse on colours is very modern, because it relies on the psychological effect of the desert landscape on the traveller.

The last quality attached to the oasis by the author is its peaceful atmosphere, as if it was a protected sanctuary: "*il y a une paix spéciale, une incomparable paix dans cette oasis non profane, que de tous côtés l'immense desert mort environne et protège*" (Loti: 95). This peace has to be compared with a general quality of Eastern people for Westerners: it provides mental peace, like the "keif"[4] of perfect moments of meditation, "[…] *le pur plaisir de savourer son existence*" (Berchet: 15). However, the East is motionless, like Islam. It is eternity according to Loti: "*l'orient éternisé dans son rêve et sa poussière*". (181). This means that somebody has to be an active agent. Somebody is moving in that peaceful and motionless space: the Western traveller. He is the one in search of the oriental source. He takes off his shoes and walks on worn rocks or sand soft as velvet: "[…] *où l'on marche comme sur du velours*" (95). Nevertheless, the sacred approach, in which the oasis is considered a temple, is contaminated or polluted by another oriental theme: the oasis is feminine. It is associated with a very common dream among travellers in the East: the harem. Pierre Loti is a passionate lover of Constantinople and the Ottoman Empire. Because of the soft texture of the rocks, and because of their pink and red colours, he imagines that the pink granite natural pools are like those found in oriental palaces: "[…] *dont le fond transparaît comme celui des artificielles piscines pour les ablutions des sultanes ou des houris*" (93). The oasis area with its natural assets has completely transformed the Western traveller into someone else: an oriental or Arabian inhabitant.

2.3 Un Été Dans Le Sahara *by Fromentin (1887)*

With Eugène Fromentin, the very famous orientalist painter and, at the same time, a renowned novelist, we turn back to Algeria, where he wrote his travel notes about the Sahara and the Sahel, in a French colonial context. The area is a very conflictual zone, even nowadays, a country at war. Despite this, and even though Fromentin suggests the negative aspects of French colonisation for the towns under French rule, he is considered an artist whose main interest lies in landscapes. He is a good example of the importance the oasis can have in a literary descriptive system applied to the desert. An empty space is not easy to describe whether you are a writer or a painter. Fromentin travelled in Algeria like so many other orientalist painters, in order to improve his works with a picturesque reality, but he wrote a very interesting diary at the same time, in which he tries to give a literary account of what he sees as a painter. So, instead of describing the monotony of dry areas, Fromentin highlights the times when the travellers stop at night, and sometimes for several days. This means that the desert is not a desert at all, at least the Sahara beginning in Algeria. We enjoy the

[4]The keif in Arabic culture is a word that gives the idea of a feeling of emptiness and fullness that can be felt when one is outside ordinary reality with the help of commonly used drugs, such as hashish. It also has a more general meaning of living to the full.

discovery of oases, but most towns are linked with gardens, green outskirts, suburbs covered with gardens, and urban greenery without which no town could survive. As we have already seen with the enchanting novelist, Pierre Loti, everything is a question of colour. The oasis causes the writer real difficulties. Moreover, in the case of Fromentin, the painter is not just eager to reproduce the outside world. What is more important is the harmony between the inside and the outside. He seeks an "*accord d'impressions*" (Mougin-Haddad-Wolting: 339). This concern, in a multi-disciplinary way, as a painter and writer, can explain the strange feeling of the reader who, through the narrator's eyes, sees life as a play performed by another (339). *Un Eté dans le Sahara* begins just like the usual travelogue. Fromentin leaves Alger for one of the most famous oases in Algeria, Biskra, at the end of February 1848.

He leaves the Tell of Constantine, one of the richest and wettest areas where agriculture is easy, at El-Kantara (it means the bridge) with 25,000 palm trees and a deep river. The painter and writer uses a lyrical tone to celebrate an early spring in what is called "*un petit village couleur d'or*" (Fromentin 1887: 5) or "*ce village en fleurs*" (5). The colours are essential: gold for the sun, green for the trees, and white for the flowers. A distinctive touch in the description is the importance of sounds (birds, a flute, a "muezzin"), and we find again the comparison with the rise of a theatre curtain when the travellers enter the desert: "[…] *cette subtile apparition de l'orient par la porte d'or d'El-Kantara*" (6). The costumes and the decor are borrowed from the Bible, especially the "[…] *pays de Chanaan, moins l'abondance*" (39). For the writer, the question of "*la couleur locale*" is very important. The words used to increase the local effect are toponyms, such as "*rass-el aïoun*", the spring head (69), and the transliteration of Arabic words ("*douar*", "*bordj*", and "*oued*").

The first big town is finally reached after having crossed the Ouled Naïl mountains. At the beginning of June 1853, Fromentin is in El-Aghouat (today Laghouat), between the "Hauts Plateaux" and the "Grand Erg Occidental", in the Saharan Atlas. From El-Aghouat, Fromentin has a view over the green palm trees, but he is first interested in another great enigma: the desert, which he tries to define. It is a heart-rending landscape: "[…] *ce tableau ardent et inanimé, composé de soleil, d'étendue et de solitude*" (121). It cannot be compared to any other. Then, the author describes the town, apparently asleep. At first, two realities mingle, because the oasis, which surrounds the town of El-Aghouat, is "[…] *aussi muette et comme endormie de même sous la pesanteur du jour*" (124). However, the oasis is more difficult to describe because it mixes contradictory qualities. It is cohesive, made of a single stuff, but at the same time, it is discontinuous, made of different pieces, such as a patchwork. The oasis of Al-Aghouat is compared to "*deux carrés de feuilles enveloppées d'un long mur*" (124). It is drawn like a park on the "*plaine stérile*". So, it is characterised as a walled, closed island, isolated from the desert, which is the most usual way to describe an oasis.

The author adds that "[…] *vue de cette hauteur, elle apparaît comme une nappe verte*" (125). He notes "[…] *quelques maigres carrés d'orge d'un jaune ardent*" (125), and he sees the desert trying to invade the gardens. Green and yellow surfaces are the oasis as seen by an observer who is a painter. The first impression is one of discontinuity. This is confirmed by the differences in colours, to which can be added the darkness of birds hidden in the trees. The desert is thus not separated from the oasis. On the contrary, it penetrates the oasis but in a way that enables the writer to introduce a comparison with a French landscape: "[…] *à chaque instant, une nouvelle trombe de poussière passait sur l'oasis et venait s'abattre sur la ville; toute la forêt de palmiers s'aplatissait alors comme un champ de blé*" (128). Later on, Fromentin compares the oasis of El-Aghouat with a "*Normandie saharienne*" (139) encircled by the desert. Still concerning El-Aghouat, the author assimilates the oasis to a town, but more probably to an Arabian architectural and urbanistic reference: the "medina". In colonial or Western novels about North Africa, the medina is often compared to a maze. The oasis is also compared to a maze. "*Malheureusement, l'oasis ressemble à la ville; elle est resserrée, compacte, sans clairières, et subdivisée à l'infini. Chaque enclos est entouré de murs, et de murs trop élevés pour que la vue s'étende de l'une dans l'autre*" (138). It is a general statement to compare the oriental town with a labyrinth, which gives an ambiguous pleasure (Berchet: 15). The Western traveller is often looking for his deepest truth in the medina (or in the oasis), and he needs to get lost in the many lanes and narrow streets to find his way, and symbolically to find himself, like the hero of André Gide in *L'Immoraliste*. Fromentin emphasises the subjective meaning of his impressions of Algeria. He says that what he wrote many years ago has no meaning except for himself and a few friends; it is a personal knowledge.

The second oasis visited by Fromentin was born because of "Oued M'zi", described as a "[…] *chemin de sable, couleur de lavande, entre deux rangées verdoyantes de roseaux et un double taillis de bois touffus* […]" (152). Its name is Tadjemout. The town is surrounded on three sides by gardens, following the path of Oued M'zi. The writer repeats his remarks about El-Aghouat, but in another way; he is obsessed with the green colour, a very special green. "*Une seule chose résiste à la consomption de ces terribles étés* […] *c'est la couleur verte des feuillages, couleur extraordinaire dont nous n'avons pas l'expression dans les harmonies ordinaires de la palette*" (166). This colour that cannot be reproduced by a painter is described with words using an original comparison between palm trees and toys: "[…] *ce badigeonnage de vert émeraude, entier, agaçant, et qui fait ressembler tous ces arbres à des joujoux de papier vert qu'on planterait sur du bois jaune*" (166). The yellow colour is introduced because the gardens at the foot of the green trees are very dry. A last quotation will prove that, finally, the writer is perfectly in tune with the painter by suggesting, in a colourful way, the beautiful landscape of the oasis gardens, or "*ksours*"[5]: "*Par-dessus tout cela, les aigrettes des palmiers d'un vert froid, légèrement jaunes ou rougissantes au point de junction des palmes, voilà les jardins de Tadjemout, c'est-à-dire de tous les k'sours du Sud*" (167).

[5]Ksours (singular: ksar) are granaries to store wheat, corn, and many food items; they are usually divided into cells called "ghorfas". They can occupy a very wide area and become real towns. Fromentin probably uses the term in the colonial sense, because French colonialists developed "ksours" to control the space.

Eugène Fromentin is a special example because the painter became a writer who wanted, above all, to express his personal feelings for the East, the Arabian world, and especially the desert landscape. He presents himself as one of the "*écrivains descriptifs*" (Fromentin 1887: VI). However, his method of expression, as a writer who describes what the painter cannot say, is based on "*l'image réfractée*" or "*souvenirs condensés*" (IX).

2.4 L'Immoraliste *by Gide (1985)*

The story of *L'Immoraliste* takes place in "189…", as it is written, and a large part of it is located in Algeria, between El-Kantara, Biskra, and Touggourt, not far from the itinerary of Fromentin. The description of the oasis is less important than the use of different elements associated with it, like the wind in the palm trees, or the paths and gardens inside the oasis, again considered a maze. The structure of the novel is more or less repetitive. Michel and Marceline first go to the Biskra oasis and stay there for one winter. Then, they take a second trip in spring, first in Italy going south until they reach Sicily, and then again to Biskra and Touggourt. They go south in order to find some heat as both of them are sick. Michel recovers from tuberculosis, but then Marceline becomes sick and dies in Touggourt after the second journey to the oasis. The oasis, a very fertile land, with the importance of the orchard theme (*le verger fertile*), is surrounded by death associated with the desert. *L'Immoraliste* is a story of rebirth for Michel and of death for Marceline, and the centre of it is in the Biskra oasis. The relationship between the French hero, Michel, and the Arabian world (or the East in a wider sense) is one of desire. The young scholar Michel desires health and initiates a sexual coming-out in the oasis. The Biskra hotel is the place where he recovers his health and meets very young Algerian boys who are like the fruits in the oasis. At the same time, this sexualised relationship takes place in a holy and religious context, associated with the New Testament, another reference to the East. The young children have a golden skin, like in the case of Gauguin, and they give Michel the taste of Antiquity, as if they were Theocritus shepherds in Sicily. As a protestant, but at the same time as a specialist of Antiquity, he is delighted when he meets the Arabian world, where he can feel a biblical atmosphere. He becomes aware of the physical beauty of these young men. His own body becomes the centre of the oasis. Health and beauty are equal in Bachir's body for example. Thus, the oasis, with its gardens and orchards, is the perfect scenery for his rebirth. His recovery from illness could not have happened elsewhere than in the oasis. The setting of their stay is perfectly beautiful. More than that, it is an Eden-like place: the scent of flowers, luminous air, turtle doves. These normal aspects of an oasis in the twentieth century are transformed into a mystical phenomenon. The palm trees and the wind swaying the palms play an important part before the appearance of a young shepherd playing the flute. The sun filtering through the palms onto Michel's face is a mystical symbol: "*je sentais le soleil ardent doucement tamisé par les palmes* […] *le vent léger dans les palmes*"

(Gide 1985: 49). This means that Michel is healed by the wind and the sun in the palms, and not only his body but also his soul is invigorated. The second time in the Biskra oasis, Michel remembers his first experience: "*Je reconnais l'immobilité des palmiers*" (179). At that time, he was reading Christ's words to Saint Peter: "*Maintenant tu te ceins toi-même et tu vas où tu veux aller; mais quand tu seras vieux, tu étendras les mains…*" (57). The stillness and silence within the gardens in the oasis are the reasons for Michel's experience. The wind passing through the palm trees is very different from what we have seen with Fromentin. Here, a spiritual presence encourages Michel to listen to the voices of his desires. This very positive experience in the oasis is unique. The second experience is one of disappointment. The oases are presented as "*peu riantes*" (179), and the narrator prefers the desert. The wind is no longer the sign of the presence of God but an image of hell and death. This wind is destructive: "*simoun ardent*" or "*sirocco aride*" (182). The experience of the oasis in Touggourt is completely negative and in tune with Marceline's agony. Finally, the stay in Algeria becomes a nightmare for Michel who suffers from the ambiguity of the East, and here of North Africa: "*Entouré de splendeur et de mort, je sens le bonheur trop présent et l'abandon à lui trop uniforme*" (185).

3 Mystical Arabic Novels

We will leave aside Bahaa Taher, *L'Oasis du couchant* (2007, translated in 2011), because we have devoted a conference, published online, to this Egyptian author among others.[6] Its scenery is the famous Siwa oasis. Two other Arabic novels, *L'Appel du Couchant*, written by the renowned Egyptian disciple of Naguib Mahfouz, Ghitani (2000), and *L'Oasis cachée*, written by the Libyan novelist, Al-Koni (1997, translated into French in 2002), will be enough to understand how far the imagined world of the oasis can go when it is rooted in pre-Islamic thought or in Islamic mysticism; the desert and its oases are part of a mystical quest.

3.1 L'Appel Du Couchant *by Ghitani (2000)*

In this novel, we have another travelogue, but it is difficult to identify the itinerary precisely and where the traveller first met the oasis. The Egyptian context influences the beginning of the story, but afterwards? A man is responsible for writing down the story of a foreigner ("*L'étranger*") who came from the East to the West

[6]Marc Kober, "*Fiction des oasis dans la littérature égyptienne contemporaine*", CRESC and PRODIG. Conference "*Oasis dans la mondialisation: ruptures et continuités*" December 2013, HAL Id: hal-01024367 https://hal.archives-ouvertes.fr/hal-01024367. Submitted 16 July 2014, pp. 13–20.

(“*le pays du couchant*”). “*L’étranger*” is an Egyptian born in Cairo. The writer explains how the journey began and how the lonely traveller decided to go west, and came across an oasis called “Oumm El-Saghîr” (*La mère du petit*) (Ghitani 2000: 67). This oasis is very strange, because it remained unknown for a very long time: “[…] *demeurée jusqu’alors inconnue des autres contrées*” (69). No precise description is given apart from palm trees and good smells; no traveller has entered the oasis for seven generations. In its eastern part, there is what is called “*Le Campement*” (the camp). The unique wealth in the oasis is the spring, “*sans pareille dans le monde habité jusqu’alors*” (71), because its temperature changes three times in a single day, and an old man called “*L’Éclaireur*” who lives under two palm trees. This man is able to cure sterility and knows meteorological changes. In these Arabic novels, including Ibrahim Al-Koni’s, the origins of the oasis are always mentioned. In this case, somebody happened to know where the spring was under the sands, and one date stone gave birth to a palm tree with a female branch and a male branch. The spring is called “*adhâra*” (“*les vierges*”/the virgins). The story of the oasis has never been written. Symmetrical to the oasis is “The camp”, but it used to be only inhabited sands. “*Le Campement est une sorte de mirage*” or “*une vibration de la lumière*” (96). It is a mental space where one can draw the figures of one’s mind. Some words can give the image of it, such as sentries (“*sentinelles*”), voices, or lines, stretching towards infinity. There is a misty atmosphere above “The camp” that hides it, and a man-made wall separates it from the oasis. The oasis cannot exist without the camp, or it is a reflection. The oasis is very uncanny. The traveller will leave after having met a woman from the oasis who wanted to become pregnant, in order to maintain the exact number of inhabitants in the oasis. The traveller goes towards the west, because the oasis was only one step in his journey and in his life. He tries to locate the oasis because he wants to come back one day. After a short walk, he enters a country where he becomes the “Prince of Deserts”, but when he talks about the oasis, nobody believes him. He finally leaves that powerful country, without being able to separate the reality from the illusion. He loses the sense of time too: “*Il flottait dans un vide sans limites*” (331). The writer understands that this old traveller was full of nostalgia and homesick because he had lost his past. This novel ends in a very enigmatic way, especially if we know that this traveller was first received personally for several days by Ibn ‘Arabi himself. A Sufi wisdom may explain the deep meaning of the whole story.

3.2 L’Oasis Cachée *by Al-Koni (2002)*

The novel is centred on an oasis called “*wâw*” (Al-Koni 2002: 15). Birds and men from the desert look after this oasis. Just like in Ghitani’s novel, a voice is calling. A hidden voice urges the nomads to meet the desert, and at the same time, “[…] *lui agitant sans cesse la promesse de l’oasis, du rendez-vous jamais accompli*” (Al-Koni: 9). This nomadic group comes with a flock of birds migrating towards the North. Twice a year, the birds leave the oases and land in the camps. They are considered a good omen; the men hear a voice calling them. They are reminded of the nomadic law: they never end their quest because they are looking for what is hidden by the large area of the desert. The nomads expect their chief to leave the camp because he can only be the poet of the oasis, "[…] *car le nomade se contente de la wâw du poème lorsqu'il découvre qu'elle est sans existence ici-bas*" (Al-Koni: 30). The whole story of Ibrahim Al-Koni is based on a very old Saharan legend, according to which the "*petite wâw*" existed somewhere in the Great Libyan Desert, where there were two other real oases, but this small *wâw* decided to appear or not to save the nomads from thirst. The small *wâw* was an image of Paradise. An old crane (*une grue*) was left behind by the tribe as a negative sign but, at the same time, the old bird has a very important value as it is an ancestor (*amgâr*). So the bird is protected by the tribe's chief against the cruel children who want to kill it. This Libyan novel is written like a poetical myth. The story is not very important compared to the different aspects of a real oasis legend. The chief finally leaves the camp and goes down into the valley. The description of the valley is a comparison with Eden, a sort of garden of beauty and innocence: "*Il était descendu dans la vallée encore vierge, tapissée dans son fond d'un sable lisse et soyeux, strié de plis charmants comme au premier matin du monde où l'ancêtre est sorti de son royaume et a marché pour la première fois dans le désert. Sur ce tapis se pressaient les genêts* […]" (Al-Koni: 33–34). Again, the hero hears the singing of the birds. The oasis is associated with youth and freshness, and it works like a Fountain of Youth, but the chief feels old and tired as if his power has vanished. The old crane dies, which is a bad omen according to him. In the "*Hamada du couchant*",[7] the nomads hope to find an escape from the persistent drought; they walk towards shallow valleys, "*miroitantes d'eau* […] *à la lisière des nuages passants*" (103). These hidden and wet valleys are all in the "Erg of Zellaf", rich in water. People from the eastern part of the hamada never visit the "hamada of the west". Apart from the chief, there is also an augur who goes to the "*hamada du couchant*" as well. This novel is very difficult to understand according to Western standards because, for example, the augur's departure is mental. The psychic forces are as important as the physical forces. We are in a magical world where all elements are double. A black kid is slaughtered as a sacrifice, and the augur has his throat cut together with a crow. It is the voice of the knife. Then, the oued flood begins and, after three days of rain, the land is green. Murder is associated with floods and blood with water in what is called "*le genie meutrier des crues*" (138). As is often the case in this poetical novel, many characters are poets, singers, and birds, and the poets sing of the beauty of the red hamada. They compare it with "*l'amante du ciel*" (194). "*Al-hamâda al-hamrâ*" (the red hamada) is at the same time precisely located and identified as "*le grand plateau pierreux à l'ouest du desert lybien*" (44). As is usual, a well is needed (*le puits de la hamada*) to create a real oasis, an old and mythical well. Some ropes and stones are used to help search for it. After the bird, the chief, and the augur, a new character comes onto the stage: a sort of supernatural road

[7]The "hamada" is a rocky plain in the desert. It is sometimes associated with a "reg".

worker called "*Le Terrassier*" who will give birth to the oasis. He explains the death of the augur and the secret meaning of earth and water: "*l'eau est l'offrande de la terre. Et rien n'a autant besoin d'offrande que le sang*" (160). He has a special relationship with earth, considered a sacrificed body. The earth is like a victim, and water is the blood of the earth according to him. Even the mysterious call from the beginning of the story finds its meaning. It is a gift from the sky but also the language of the earth (166). The story of Ibrahim Al-Koni is obviously based on a poetical and metaphysical approach of geographical realities of the desert area. The history is presented in a mythical way since the origin of the oasis, its foundation, relies on the art of digging a well. The "*Terrassier*" and his only son discover wet clay. Then, the "*Terrassier*" disappears in the well, because his blood is the price for the water in the oasis. It is the mythical birth of the oasis called "*la petite wâw*" (171). Inside the small valley, the broom is green again and its perfume can be smelt all over the desert, but the author then imagines an endless forest. Once the natural and luxuriant green setting is there, the author imagines the human settlement: "*l'oasis se peupla d'artisans, de marchands, de filles et des vanniers*" (176). The oasis will be called "*petite wâw*" or "*Tân Amgâr*" ("*la terre du chef*") and that is the way the author explains the mythical origins of this hidden oasis.

In the case of a storytelling, the origins of an oasis, a foundation story, like that of Ibrahim Al-Koni, or Gamal Ghitani's novel, the naming of the oasis is very important. In fact, it represents, in the original language, the story of its origin, exactly related to the local reality and to the mental background of the people living there. For writers from the West, the question of the origin is not so important and, most of the time, they just ignore the legendary context. They are less sensitive to the internal qualities than to the external appearance of the oasis, which can easily be related to the Garden of Eden and the Bible. This idea of a pure area, preserved from any kind of pollution, which would help the desert inhabitants (or those who only cross it, such as nomads or travellers) to survive, is not reserved to Europeans. When it comes to the question of how to describe an oasis, there are many common elements, such as the different types of oasis according to the origin of the water and vegetation, the proximity of a town or not, and the organisation of gardens and orchards within the oasis. In the case of writers who consider literary description an art, such as Pierre Loti, André Gide, or Eugène Fromentin, we can feel the influence of a common ideal at the end of the nineteenth century, the "*écriture artiste*", the importance of colours and painting, the priority given to art, but also a physical experience, a very deep change in the body and mind that is part of the oriental, colonial, or exotic experience.

References

Al-Koni I (2002) L'Oasis cachée [Wāw al-ṣuġrā]. Phébus, Paris (Translated from Arabic by Philippe Vigreux)

Bahaa T (2007) Oasis du couchant. NRF/Gallimard, Paris

Berchet J-C (2001) Le Voyage en orient. Robert Laffont, Paris
Cogez G (2014) Partir pour écrire. Figures du voyage. Honoré Champion, Paris
Collot M (2014) Pour une géographie littéraire. José Corti, Paris
Elmarsafy Z (2012) Sufism in the contemporary Arabic novel, Edinburgh studies in modern Arabic literature. Edinburgh
Engelibert J-P (1997) La Postérité de Robinson Crusoé: un mythe littéraire de la modernité (1954-1986). Droz, Geneva
Fromentin E (1887) 1857 Sahara et Sahel. 1. Un Eté dans le Sahara. 2. Une Année dans le Sahel, Plon, Paris
Ghitani G (2000) L'Appel du couchant [Hātif al-mag°īb]. Seuil, Paris (Translated from Arabic by Valérie Creusot)
Gide A (1985) 1902 L'Immoraliste. Folio, Gallimard, Paris
Loti P (1913) Le Désert, 58th edn. Calmann-Levy, Paris
Mougin P Haddad-Wotling K (2002) Dictionnaire mondial des littératures. Larousse, Paris
Sussman P (2010) L'Oasis secrète. Pocket, Paris
Wallut C (1876) Grandeur et décadence d'une oasis. Albanel et Baltenweck, Paris
Wallut C (1883) L'Oasis – scènes du désert. C. Delagrave, Paris

The Conceptual Approach of Oasis as Insights on Globalisation. Example of the Coastal Valleys of Northern and Central Peru

Evelyne Mesclier, Anaïs Marshall, Célia Auquier and Jean-Louis Chaléard

Abstract Beyond the description of a specific type of landscape, the concept of an oasis is useful to understand the spatial transformations produced by contemporary globalisation. The coastal valleys of northern and central Peru are a good example. The logic of their changes is based on the assumption of the greater competitivity of large agricultural companies, which has led the actors, including the State, to reorganise the space in a particular way. Today, territorial enclaves adjoin the old irrigated areas. This chapter questions the issues of this reorganisation on the basis of two characteristics specific to oases: their organisation around the circulation of water and their role as crossroads of human movements, which is often reflected by the constitution of true centralities. This conceptualisation of the oasis, no longer as a type of landscape but as a system of spatial relationships, enables an analysis of the way in which the enclaves reorganise, paradoxically, the whole of the spaces in which they develop. This process seems to be a relevant example of the transformation of territorial hierarchies in globalisation.

Keywords Enclave · Crossroads · Globalisation · Oasis · Periphery · Peru

E. Mesclier (✉)
IRD, UMR PRODIG, 2, rue valette, 75005 Paris, France
e-mail: Evelyne.mesclier@ird.fr

A. Marshall
University Paris 13-Nord, EA PLEIADE, UMR PRODIG, Paris, France

C. Auquier · J.-L. Chaléard
University Paris 1—Panthéon-Sorbonne, UMR PRODIG, Paris, France

E. Lavie and A. Marshall (eds.), *Oases and Globalisation*,
Springer Geography, DOI 10.1007/978-3-319-50749-1_3

1 Introduction

The term oasis primarily brings to mind those found in North Africa or the Middle-East: isolated green areas in the middle of the desert with terraced crops under the date palms. Nevertheless, it is also used in other regions of the world and in other climates, such as in China, the USA and Australia, and for other spatial forms, for example the piedmont valleys. The term may be so common because it has become a conceptual tool, beyond the object that it originally described. In fact, it is no longer only used by geographers to describe the locations, modes of access to water and forms of land around agricultural crops specific to desert areas, but also to analyse the systems of physical and social constraints that characterise the settlement of people in hostile environments. It is also used to understand the type of multiscale spatial organisation that explains the existence of the oasis, around the idea of a meeting place at crossroads (Retaillé 2006: 672).

We will use this double concept, of social organisational constraints to manage the circulation of water, on one hand, and crossroads and a centre that results from it, on the other hand, to analyse the changes that have transformed most of the oases of the Peruvian coastal piedmont, particularly in the north and centre of the country. These intensely cultivated and extremely urbanised valleys in the middle of the desert have apparently lost their importance in a coastal region increasingly dominated by agro-industrial firms situated in the interfluves. Although they have led to the constitution of enclaves at first sight independent of the old valleys, these transformations, paradoxically, have resulted in significant disruption of both the management of water resources and the economic, social and political organisation of local societies. This can be better identified by the concept of oases.

First, we will show how the cultivated and urbanised valleys of the Peruvian coastal piedmont represent true oases, both very dependent on the concerted management of water resources in a partly hostile environment and central to the spatial organisation of their regions and national territory. We will then see how the current transformations of the landscapes reveal not only a simple expansion of these oases, but also the appearance of a new logic that falls within contemporary globalisation. Lastly, we will deal with the changes in spatial relationships produced by this new logic.

2 Cultivated and Urbanised Valleys in the Desert

The coastal oases of northern and central Peru have been intensely cultivated and densely populated for centuries. Until two decades ago, their functioning relied on the ability of societies to ensure the circulation of resources and products both along the coast, on a north–south axis, and along the valleys linking the slopes, the piedmont and the sea, on an east–west axis.

2.1 Contrasting Landscapes between the Sea and the Cordillera

The western piedmont of the Andes (Fig. 1) is considered a desert from the north of Peru to the centre of Chile, between approximately 5° and 27° of latitude south: "*a desert because it is anticyclonic, coastal because it is blocked by an orographic obstacle, cold because an upwelling runs parallel to it*" (Demangeot and Bernus 2001: 18)—J. Dresch describing the role of the latter in the aridity of the environment (1982: 34). This piedmont is interspersed by around fifty rivers, of varying flows and marked seasonal variations, which descend from the cordillera over relatively short distances: sometimes there is only about a hundred kilometres between the source and the mouth (Dollfus 1968: 135 and 143). The valley heads are situated at altitudes varying between 2500 and 4500 m and dominated by summits culminating at around 4000–6000 m depending on the latitude. The western slope of the cordillera is narrow and abrupt, the gradients are steep and the rivers are enclosed in the middle part of the valleys. In their lower part, at around 500 m of altitude, these gradually widen towards the alluvial fans of the coast (Pulgar Vidal 1940; Dollfus 1968). The generally flat areas of the coastal strip are interspersed by low-altitude reliefs; coastal mounds in the south, rocky hills in the centre and horsts in the north (Dollfus 1968: 138–139).

The location of Peru in an intertropical zone explains the seasonal nature of precipitation on the cordillera that dominates the piedmont. It is greatest during the summer months, essentially between November and April, when the rivers reach their maximum flow. The rest of the year, the coastal rivers, have very little water and some of them are completely dry. There is great interannual variability, with prolonged droughts and phases of very heavy rainfall, particularly during El Niño events, which occur about once every ten years. In general, and at all temporal scales, their flow is "subject to sudden pulses" (Collin Delavaud 1968: 67). In addition, the composition of the alluvial fans of the piedmont results in the accumulation and circulation of underground waters, in the form of superficial and deep layers (*ibid.* 71).

While seasonal precipitation is almost absent on the piedmont and very restricted in the southern cordillera of this whole area, it is more abundant in the centre and especially in the north of Peru, where the influence of the upwelling gradually fades. In the south of Peru, the "*(…) Pacific desert climbs high: the "arid diagonal" hits the western slope of the Andes sideways on*" (Dollfus 1968: 33). The precipitation that falls on the Pacific slope of the Andes gradually increases towards the equator. In the north of the country, the mountains are "greener", "(…) *the desert only touches the foot of the mountain*" (*ibid.* 92). In addition, due to its geological history, the piedmont is considerably wider in the north of Peru, where it can reach around a hundred kilometres, whereas in the south of the country the valleys only slightly widen on arriving at the sea (Dollfus 1968: 166–167).

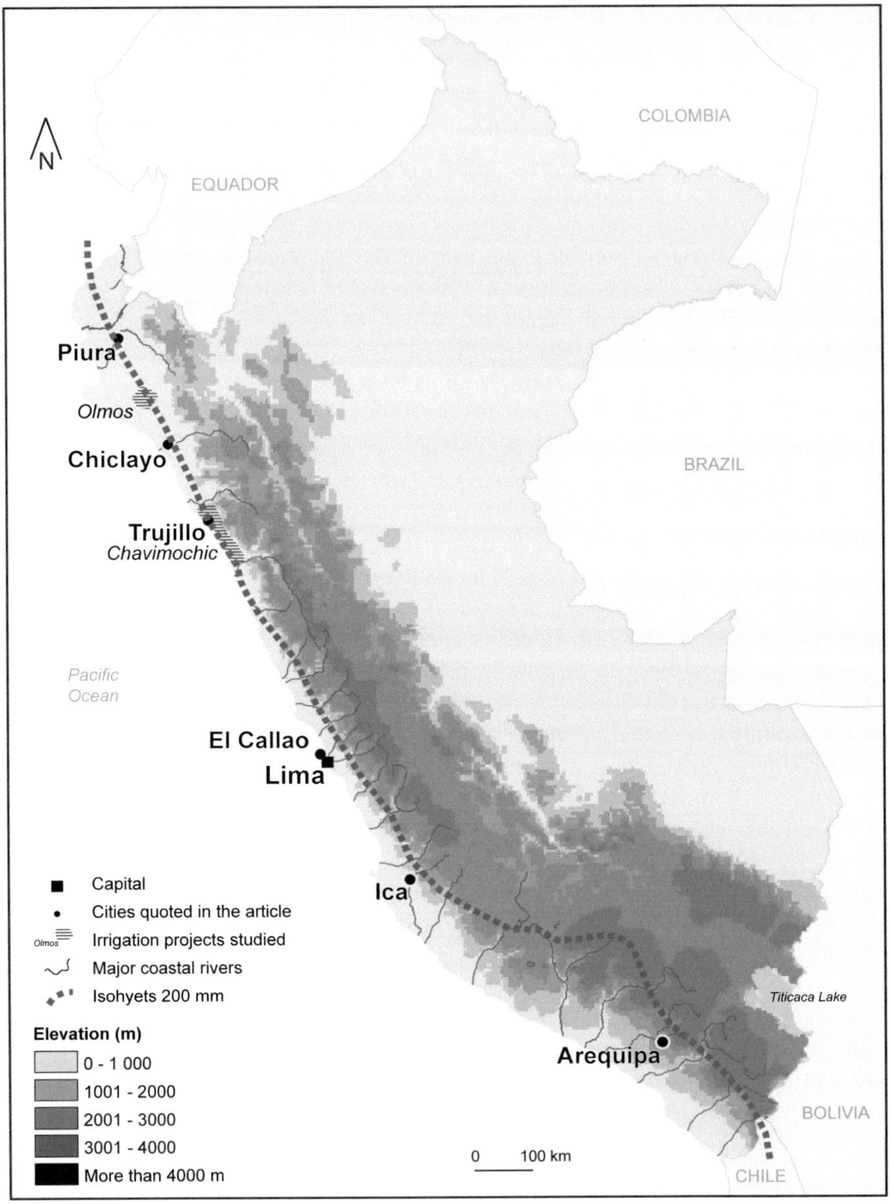

Fig. 1 Location of the mentioned areas. *Sources* IGN Peru, 2005; ANA, s.d.; Google Earth 2016; Field work

The agricultural landscapes of the northern and central coastal valleys, observed from a high point, recall those of African oases. In the north, "*on the piedmont itself, the coastal rivers, denser and better supplied at these low latitudes, have constructed many great alluvial plains that form as many oases, vast and well-watered, narrow bright green tracks infiltrating the heart of the ranges with their bare, brown and reddish walls, or fans fully open to the Pacific where the green checkerboards butt directly against the purple and ochre rocky desert*" (Collin Delavaud 1968: 17). At Ica, in the centre of the country (cf. Fig. 1), the landscape "*resembles an oasis: an intense green strip of vegetation, situated in the middle of the desert and surrounded by impressive sand reliefs that are commonly called "dunes"*" (Oré 2005: 56).[1]

2.2 *An Intensive Commercial Agriculture Organised around the Management of Water*

At the mouth of the valleys, the northern and central alluvial plains have enabled the development of vast irrigated areas. Since the end of the nineteenth century, the societies of the coastal piedmont have accessed the underground water via excavations, tunnels, then wells and motor pumps. They have also used the surface water via canal systems, which from the pre-Hispanic period could link several valleys (Oré 2005: 42). These areas led to the possibility of a dense population and the need for the concerted management of the distribution of water rights and the maintenance of canals, which for centuries have involved the populations of both the valleys and the Andean slopes (Rostworoski de Diez Canseco 1981: 23).

From the sixteenth century, and despite part of the irrigation infrastructure being abandoned or damaged during the conquest, the Spanish developed Mediterranean crops in these valleys, thanks to the mild temperature all year round: vines in Ica, sugar cane around Trujillo, extensive sheep and goat farming around Piura (Piel 1975: 152–153; Schlüpmann 1994: 125 and the following). These goods were destined for both the regional and city markets. After independence and the creation of Peru as a nation state, and especially at the end of the nineteenth century and the beginning of the twentieth century, export crops became increasingly important in the national economy, with the progress in land and sea transport and the expansion of European and North American markets. Large domains appeared based on cotton and sugar. Peruvian traders and foreign investors bought up the private haciendas in decline and the church-owned lands, as well as the small farms and village community lands (Klarén 1970: 41 and the following; Huetz-de-Lemps and Collin Delavaud 1983: 44). Vast domains were formed, sometimes covering several thousands of hectares but with a single tenant.[2]

This model based on great land ownership was radically changed by the land reform of 1969, but without the intensity of the land occupation decreasing. The

[1]Translation from Spanish into French by the authors of this chapter.

[2]Like *Casagrande* near the town of Trujillo (department of La Libertad) or *Cayalti* near the town of Chiclayo (department of Lambayeque).

large sugar-producing domains and some of the more modest properties were transformed into cooperatives by the State. Some entered a phase of decline due to both management and market problems, as world prices dropped dramatically at the end of the 1970s (Huetz de Lemps 1983: 83). Nevertheless, the growth in the domestic demand for sugar and cotton provided new outlets and several of the sugar-producing cooperatives were still in full production at the time of the privatisation implemented by the State in the mid-1990s (Revesz 1991; Mesclier 2000; Chaléard et al. 2008). The other cooperatives were often disbanded by their members, who formed their own farms. The urban transition that accompanied the demographic growth of Peru did not depopulate the countryside. In the 1980s, although the population growth rate in the piedmont countryside was negative within a radius of more than 200 km south and north of Lima and near Piura, it remained positive almost everywhere else (Huerta et al. 1997: 47).

2.3 Central Spaces Organising the North–South and West–East Movements

These green and intensely cultivated valleys have played a growing role in the relationships between regional spaces and the exterior. The Spanish founded towns there at the beginning of colonisation in the sixteenth century. These are often situated slightly back from the ocean, connected to a port or a quay (Le Callao for Lima, Paita for Piura, Salaverry for Trujillo) on one side, and to the middle and high part of the valleys on the other, and linked to each other by a road that became the Pan-American road in the twentieth century (Deler 1991: 282).

The towns took advantage of the agricultural expansion in the twentieth century and gathered together the administrative, commercial and service functions as well as welcoming the great landowners. During the urban transition that accompanied the demographic transition of the twentieth century, mainly from the 1950s, they experienced an exponential growth: this is particularly true of Lima, the administrative, political and economic capital of the country, which today has 9 million inhabitants. It is also true of the main towns of the piedmont, in which the mean population growth rate in the 1980s was almost always higher than 2.5% (Huerta et al. 1997: 49). The population of Peru was increasingly concentrated on the coastal piedmont during the twentieth century: according to the INEI, in 1940 fewer than 1.8 of 6.2 million Peruvians lived in the coastal region, i.e. 28%; in 1993, it was 11.5 million, i.e. more than half of the national population (INEI 1995).

The great sugar-producing domains of the twentieth century also contributed to this urban concentration and this crossroads of movements in the oases. They built railways that carried their products from their lands to the ports. They also participated in linking the piedmont and the slopes, to take advantage of the complementary ecology of the different altitudinal stages: in the middle and high part of the valleys, they produced some of the food needed by their workers (Collin

Delavaud 1968). Over all the piedmont, the agricultural day workers came, and still come, from the middle or high parts of the valleys. Trade links were also forged between their regions of origin and the piedmont (Gómez Cumpa and Bazán Alfaro 1989; Chaléard 2010).

The creation of the great domains led to workers moving closer to the factory, to the extent that small towns were formed next to the cultivated land, subject to the authority of the landowners. This type of enclave was nevertheless limited by the need, for the great domains then for the cooperatives, to negotiate the distribution of water with the neighbouring farms, on one hand, and with the towns, on the other hand. Until the 1980s, the growth dynamics of agricultural and commercial activities in the countryside and in the towns remained localised in the irrigated areas situated at the mouth of the Andean valleys, and vast stretches of desert still separated the oases from each other. Since then, this landscape has changed considerably and, with it, the logic that governs the relationships between towns and the countryside and between farms.

3 The Intensification of the Logic of Globalised Enclaves in the 1990s

The extension of the irrigated areas of the coastal piedmont has varied over time, depending on the construction and maintenance of hydraulic structures and, particularly since the 1960s, with the improvement in techniques enabling water to be drawn deeper from the water table. Nevertheless, the current transformations differ from earlier variations in their rapidity and in the spaces produced, both enclosed at the local scale and directly connected to global networks.

3.1 The Accelerated Expansion of Irrigated Areas since the 1990s

Since the 1990s, the coastal region has undergone significant changes that have considerably altered the landscape. In the north and centre of the country, along the Pan-American road, it is no longer rare to travel from one valley to another through a continuous zone of vast residential areas and large agricultural properties. The city of Lima is a special case, as today the agglomeration stretches over three valleys, without interruption. The last agricultural plots have disappeared from the lower parts, previously owned by the haciendas then the cooperatives and lastly the small independent farmers, while the arid slopes are covered with residential areas that climb the slopes up to several hundreds of metres above the sea (Matos Mar 2004; Piron et al. 2015). Outside the capital, both urban neighbourhoods and vast agricultural plots have taken over the desert. For example, in the region of Virú, south

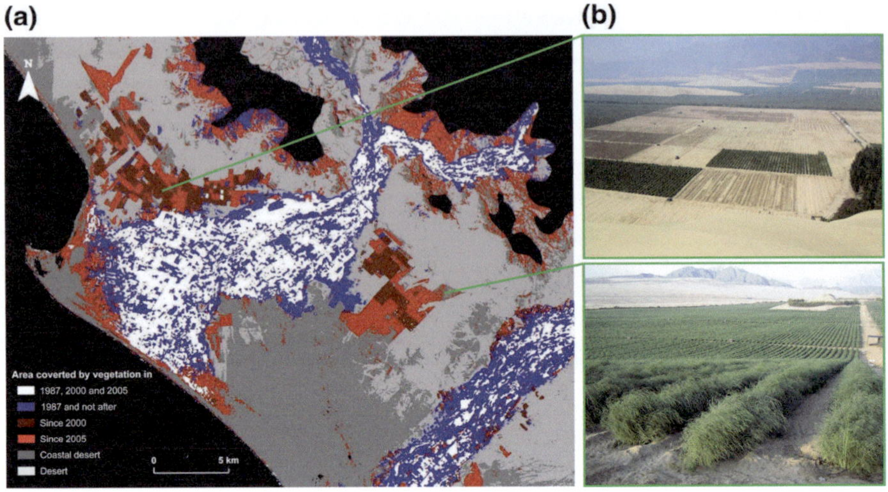

Fig. 2 **a** Map of changes in vegetation cover between 1987, 2000 and 2005 in the sector of Virú. *Sources* Image Landsat 1987 and 2000 and Spot 2005 and **b** photographs of asparagus fields in the previously desert interfluves (Pictures taken by Marshall A. in 2006 and in 2007)

of the town of Trujillo, cultivated lands are today very common outside the irrigated areas of the end of the 1980s.

In Fig. 2a, the red and maroon colours show the extension of vegetation into the previously desert zones. These dynamics are visible in two sectors: in the lower part where halophilic vegetation is developing between the coast and the plots of the old valley, and in the interfluves where large geometric plots already cover a considerable area. The desert is transformed into a green and yellow checkerboard, which changes according to the growth stages of monocultures, such as asparagus (cf. Fig. 2b).

The change in the landscape of the piedmont is due to a political desire to extend the agricultural areas beyond the lands previously cultivated and shared between thousands of small farmers. After the 1980s, marked by hyperinflation, the structural adjustment was accompanied by a radical change in the economic model (Gonzales de Olarte and Samamé 1991). Within this framework, successive governments no longer promoted small-scale agriculture, resulting from the land reform of 1969, or cooperatives, but the export agro-industry (Mesclier et al. 2013).

The interfluve lands often belonged to agricultural groups: cooperatives, associations or "peasant communities" (agricultural institutions, sometimes ancient in origin sometimes more recent, whose members collectively own vast stretches of land). The families often used the resources found there (pasture, wood from dry forests) without personalising their land rights. In the 1990s, the new land legislation increased the ways of transferring these lands to private owners, within the framework of the transformation of Peruvian agriculture (Del Castillo 1997; Mesclier 2009).

3.2 *The Logic of Enclaves at the Local Scale*

The new irrigated areas are conceived and operated in the logic of acquiring water resources outside the old valleys, whether it is in the context of large national irrigation projects or private Irrigation projects.

In the first case, the State has built significant hydraulic infrastructures, usually designed to produce electricity as well. Water is diverted to the interfluves from upstream of the nearest river, from a further but more abundant or less used water course, or from the Amazonian slope of the Andes. It is carried separately from the water of the old valleys, in specifically designated canals. This is particularly the case for the Chavimochic project. The water of the River Santa has been diverted to the interfluves situated between the valleys of Chao, Virú, Moche and Chicama via an irrigation canal more than 200 km long. The State planned to extend the agricultural land over 78,000 ha (PECH 2010). During the development of these irrigated areas, new associations or user commissions were set up alongside those that previously existed. Thus, in the case of Chavimochic, an association of users of pressurised irrigation (*Junta de Usuarios de Riego Presurizado,* JURP) was founded on 3 March 2004; in 2015, it consisted of 55 companies (according to the Internet site of the JURP Moche Virú Chao).

Further north, in 2014 the Olmos irrigation project enabled the extension of the agricultural boundary over more than 20,000 ha using water transferred from the Amazonian slope of the Andes to the piedmont. The new irrigated area is situated between two valleys that are cultivated due to gravity and pumped irrigation systems of underground water, along the Cascajal and Olmos rivers. The companies in this area buy the water according to quotas, which depend on the quantity of land bought. They also sink wells to complete their allocation.[3] They are not forced to coordinate their access to water, but they are still united by another issue—managing the risk of flooding caused by the El Niño phenomenon. They have established a common investment fund for the potential construction of drains, the protection of crops and the purchase of equipment. The producers from the old valleys have not been invited to join this association.

In a second case, the private companies, individually and spontaneously, have developed their agricultural activity by tapping the underground water and installing their own irrigation network. These companies only use water from the water table, as the surface water is too variable over the year for the continuous growth of crops. In the region of Ica (cf. Fig. 1), for example, the companies of the new irrigated areas have access to water while remaining independent of the collective management institutions of the oases; the water comes directly from private wells.

The new landowners are mostly strangers to the old valleys. Auctioning the land in very large lots (from several hundreds to several thousands of hectares) favoured

[3]According to the fieldwork we carried out between 2002 and 2015, thanks to research funds from the French Institute for Andean Studies (IFEA), the Institute of Research for Development (IRD) and The French National Research Agency (ANR).

investors with a great deal of capital. These include some of the biggest Peruvian companies, such as Gloria S.A. or Camposol, joint ventures of Peruvian and foreign capital, such as Danper, and foreign transnational companies, such as the Colombian company Manuelita S.A. The owners do not live in the oases, and their companies do not necessarily export their products via the towns of the neighbouring valleys or their ports. The products are packaged on the plots then transported by road to one of the country's major ports or the international airport of Lima.

In the landscape itself, the demarcation between these spaces and the rest of the territory is very visible: the hedges of thorny trees that separate the lands of the companies on the Pan-American road or public highways, and the gates and security guard posts that often mark the entrances, further reinforce the impression of their isolation. However, these enclaves at the local scale are directly connected to the world markets.

3.3 *The Strengthening of Direct Ties with the Rest of the World*

Whereas the companies of the interfluves seem to be autonomous in relation to the local environment, they have close links to the world markets. They use foreign capital and essentially produce for external markets (Mesclier et al. 2013). They particularly meet the new demand for out-of-season products, such as asparagus, or those that were previously not widely consumed in Northern markets, such as avocados or mangos. Thus, in 2010, in the irrigated area of Chavimochic, the main crops were asparagus (more than 50%), avocados (23%) and then sugar cane (15%), destined more for the national biofuel and sugar markets (PECH 2010: 54).

Almost all the inputs, bought in large quantities, come from foreign companies that provide improved, high-yield seeds, developed in high-tech laboratories, and chemicals certified to international standards. Irrigation is also largely carried out by imported equipment, including pumps, spray or drip systems and all the programming equipment.[4]

The companies then transform and package their product as required by their international clients, thus controlling several segments of the process. They may simply sort and clean the products that will be sold fresh (asparagus, avocados, white onions, mangos and table grapes), or pack the fruits and vegetables, especially asparagus, in glass jars or cans after cooking. Chilli peppers may be transformed into powder or canned. The products are exported, just-in-time, to the USA, Asia or Europe. They meet the stringent health demands imposed by the

[4]According to the fieldwork we carried out between 2002 and 2015, thanks to research funds from the French Institute for Andean Studies (IFEA), the Institute of Research for Development (IRD) and The French National Research Agency (ANR).

indispensable export licenses, such as the Global Gap certification, HACCP (for Europe) and Tesco Nurture 10 (for the English chain of supermarkets of the same name) (see Footnote 4).

According to all these elements, the interfluves where the companies are concentrated present the characteristics of enclosed spaces, both internationalised and disconnected from their near environment as much in their relationship to other local actors as in their relationship to water. However, this disconnection is only apparent.

4 Behind the Logic of the Enclave, a Radical Change in Spatial Relationships

Although apparently separated from these oases, the enclosed spaces created by globalisation actually have many links with them. Their presence transforms the organisation of the space much more than the just appearance of the territories of the new companies, but as much in the management and distribution of water resources as in the definition of economic, administrative and political centralities.

4.1 Water Systems Increasingly Dependent on the Enclaves

The new irrigated areas remove and put back the water in the old valleys, environments that are extremely sensitive to the circulation of water resources. The agricultural companies are at the origin of the problems of scarcity or flooding and more capable than the producers and the local authorities of the old valleys to influence their resolution.

At Ica, the increasing scarcity of water is the most remarkable. The many faults in this piedmont zone marked by a constant tectonic activity facilitate the transfer of water between the water tables. The companies established on the interfluve, by pumping the water, lower the level of these tables not only under their plots, but also under the old valley. The producers in the valley, who do not have the means to bore wells as deep as those of the companies, thus no longer have access to the underground water and become dependent on surface water, which is very irregular. Many agricultural areas are drying out in the middle of the oasis (Marshall 2014b). Paradoxically the intervention of the State to protect the water tables has made the problem of the oasis farmers worse because it has incited the companies to adopt new strategies. Boring new wells in the old valley has been forbidden since the 1970s (Oré et al. 2012). In 2005, this interdiction was extended to the whole region, including the interfluves.[5] Thus, in order to continue to increase their exploitation of water, the large agricultural companies have bought wells in the old valley and installed a network of underground pipes and covered canals to carry the water to their farms. Today, there is a lack of water for human consumption in some places (Marshall 2014b). Yet, the agricultural companies have obtained permission from the State to divert the water from a neighbouring coastal river to replenish the water table of the interfluve, without taking into account the specific issue of the old valley.

Excess water threatens the agricultural plots in the old valleys concerned by the Chavimochic project. In addition to the drainage systems of the interfluves ending up in these valleys, some of the water from the project is also deliberately discharged there, thus providing permanent access to water. The farmers can grow products for export, under contracts signed with the companies, and the latter can rent land in the old valleys, depending on the soil characteristics that they judge suitable for their activities. However, the level of the water table has risen considerably (Vasquez 2000) and, in the lower parts of these valleys, the soil is water logged. In the valley of Virú, near Trujillo, the flooding of the land has forced some farmers to abandon their plots (Marshall 2014b). In addition, this increases the naturally present salination of the soil. Paradoxically, the agricultural companies of Chavimochic have obtained some of the recovered drained plots from the project, even though these belong to the members of an association of small producers (Marshall 2014b: 182–183).

At Olmos, the water transferred from the Amazonian slope runs into the bed of one of the two local rivers which, before the inauguration of the project in 2014, flowed very rarely, "*and not even every year*" (Collin Delavaud 1968: 415) except during El Niño phenomena (Auquier 2013). The water then joins a wide concrete canal leading to a new irrigated area. The passage of the water in the valley refills the water tables situated next to the water course. Moreover, a small part of the water carried by the canal should, according to the plans of the irrigation project, be sold to the farmers in the areas closest to its path. This creates new short-term prospects for the farmers but, according to the project technicians, risks bringing salination problems to the valley in the medium term.

4.2 Centres in an Increasingly Dominated Position

The presence of irrigated areas also transforms the oasis towns. The interfluves are entirely occupied by agricultural plots. Thus, most of the new activities generated by the presence of the companies and the populations attracted by these dynamics must find their place in the old oases. Unlike in the haciendas of the twentieth century, the agricultural workforce from other regions is not usually housed by the

[5] RA n°103-2005-GORE-DRAG-I/ATDRI.

companies. Unplanned housing has appeared almost everywhere close to the areas irrigated by the Chavimochic project. For example, the town of Víctor Raúl, near Trujillo, today occupies 65 ha, whereas it did not exist in 1987 (Marshall 2009). In the region of Ica, the town of Nueva Esperanza appeared in May 2005: its inhabitants took illegal possession of an agricultural plot; in 2009, it had grown considerably and consisted of 400 families, i.e. about 1800 people. Similarly, new neighbourhoods have become established on the slopes around the small town of Olmos. These populations need not only housing, but also education and health infrastructures and means of transport other than the shuttle bus to work organised by the agricultural companies.

In the small- and medium-sized towns, the installation of bank branches, new health services, agricultural services, the growth of hotels, and the expansion of markets of agricultural products and basic needs are some of the rapid changes linked wholly or partially to the appearance of new irrigated areas. These transformations establish these centres in the new networks at the national and international scale and potentially strengthen their role as crossroads of services at the local scale.

The common point of these transformations, which may sometimes seem favourable to the development of towns, is that they are controlled not by the urban authorities, but also by the entities called "Special Projects". These manage the hydraulic infrastructures and the land transfers in the new irrigated areas, the forced displacement of populations that they judge necessary for the construction of the latter, and the aid programmes to small farmers next to these projects. They coordinate, with the Ministry of Housing and the local municipalities, the choice of plots to urbanise and, in the case of Olmos, the construction of a "new town" for the workers. These Special Projects were transferred by the State to the regional elected governments in the framework of the decentralisation begun in the 2000s. Nevertheless, in practice, they have kept a technical, economic, financial and administrative independence as well as separate premises from those of regional governments.[6]

The agricultural companies of the interfluves sometimes intervene very directly in the urban issues. Thus, the agricultural plots that they have gained in the desert now surround the town of Trujillo, forcing the town authorities to negotiate their urban expansion policies with them (Marshall 2014a). In addition, in Virú, the agricultural companies have organised a private militia to deal with delinquency and ensure their and their workers security (Marshall 2009). In the case of Olmos, the "new town" that will be built on the edge of the irrigated area also risks being managed more by the companies that will employ its population than by a municipality whose headquarters is far away.

[6]According to the sites of the Olmos-Tinajones Special Project: http://www.peot.gob.pe/ and the Chavimochic Special Project http://www.chavimochic.gob.pe/, and our fieldwork.

4.3 A Visible Refocusing of Spatial Relationships Around the Globalised Economy

The influence of actors in the global sphere of action in the organisation of spaces is here particularly visible: they have settled on the periphery of the old centres by processes that can be both traced and dated and are thus easy to identify, in all their difference, compared to the actors who have been there for much longer and whose sphere of action is local or regional. Their links with the circulation of natural resources, such as water, and with the reorganisation of decision-making powers are also identifiable.

An example of the particular visibility of the interaction of the new actors with the local societies concerns the agricultural systems. Throughout the world, it has become common for farming families, without much land, to work directly with the big agro-industrial companies, under different types of contract. This enables these companies to adjust their supply of raw materials to the changing needs of the markets. In the case of the oases of the Peruvian coast, as well as this dependence on the agribusiness sector, which often concerns only a part of their farm, the small producers are also affected by the industry's modification of the characteristics of their water supply. The influence of the agro-industry on the neighbouring small-scale agriculture is thus clearly marked in the landscape, as well as in the circulation of water and in the decisions taken regarding its management.

Those responsible for transforming the periphery are also at the origin of changes in the urban centres, due to the development of their activities. Some agricultural company managers take full responsibility for the development of the whole territory constituted by the valleys and the interfluves by standing for election as the head of the regional governments created by decentralisation.[7]

In contrast, the institutional actors established for much longer in the old oases seem to have little control over the transformations occurring in the periphery. The agricultural institutions have lost access to some of the lands they owned and the authorities devolved from the State compete with the companies to assure the functions of the police in the towns. The authorities decentralised at the regional or municipal level, if they have no other source of power than their functions, accept rather than propose the operations of relocation or concentration of populations of the interfluves.

This is not the same scenario as described in the 1980s by A. Reynaud, i.e. the formation of a hypercentre by the annexation of the periphery to an older centre (Reynaud 1983: 589). It is neither the substitution of one center for another, because the enclaves of interfluves are far from having all the characteristics of centrality: organised exclusively around economic activity, in their current state, they do not offer densely populated living areas, or public spaces, or cultural or sporting activities, or even various sites of consumption of material goods—unlike the

[7]The curriculum vitae of the candidates for regional elections are accessible on the site http://www.infogob.com.pe.

towns, or even the villages of the haciendas of the twentieth century. Power and activities are unequally distributed between two very different locations. The oases of the Peruvian coast thus appear to be a good example in which to observe the creation of new types of spaces, in the sense of systems of relationships between places, specific to contemporary globalisation.

5 Conclusions

The concept of an oasis, viewed above all as a particular system of spatial relationships, makes the current dynamics especially visible. Today, the transformations of these oases are excessively determined by the presence of agro-industrial companies in the interfluves, which reconstruct the whole of the territorial relationships to their advantage. One of the characteristics of contemporary globalisation can be observed here, marked by the "glocalisation" of these big transnational economic actors (Swyngedouw 1997). This glocalisation determines both the new territories of water, including outside the new irrigated areas, and the new social and political territories, including the towns. The enclave functions like a system of isolation between the outside and the company, but does not prevent the latter from using the oasis. The oasis, both a human settlement in a hostile environment and crossroads of relationships, supplies the agro-industry with an anchorage, certain aspects of which it exploits, like the ability to accommodate populations and services or to absorb a scarcity or an excess of water, but whose concerted management of the territory it destroys.

The oases thus provide particularly visible examples of types of current globalisation, which focus decision-making powers on the great economic actors, at the expense of democracy, if by this one means taking into account the aspirations of everyone as different aspects of human existence. The example of oases encourages globalisation to be revisited, based on systems of physical and institutional relationships especially visible here, and which may prove more difficult to update in other spatial contexts.

References

ANA—Autoridad Nacional del Agua (s.d) Elaboración de mapas de isohietas – Ámbito político, administrativo y unidades hidrográficas. Oficina del Sistema Nacional de Información de Recursos Hídricos Ms. Sc. Sigfredo Ernesto Fonseca Practicante Pamela Reategui Medina. 55 p. http://fr.calameo.com/books/001271945f0850843975e

Auquier C (2013) Transformations agraires par le prisme de la mondialisation. Le cas d'Olmos, une oasis dans le nord du Pérou. Mémoire d'ingénieur Istom, UMR PRODIG, 120 p

Chaléard J-L, Huamantinco A, Mesclier É (2008) Les dernières grandes coopératives péruviennes à l'épreuve de la privatisation. Autrepart, n° 46, pp 13–29

Chaléard J-L (2010) Les oasis de la côte nord du Pérou: des îles jusqu'où? In Comme un parfum d'îles. Hommage à C. Huetz de Lemps, Paris, PUPS, pp 455–466. <hal-00748816>
Collin Delavaud C (1968) Les régions côtières du Pérou septentrional; occupation du sol, aménagement régional. Lima: Institut français d'études andines, 600 p
Deler J-P (1991) Le Pérou entre deux mondes (ch. 22). Amérique latine. In Brunet R. (dir) Géographie Universelle. Belin—Reclus. Paris, 480 p (pp 278–295)
Del Castillo L (1997) Propiedad rural, titulación de tierras y propiedad comunal. Debate Agrario 26, Lima, CEPES: 59–79
Demangeot J, Bernus E (2001) Les milieux désertiques. Paris, Armand Collin (coll U géographie). 295 p
Dollfus O (1968) Le Pérou. Introduction géographique à l'étude du développement. Paris, IHEAL, 355 p
Dresch J (1982) Géographie des régions arides. Paris: PUF (Coll. Le Géographe), 277 p
Gómez Cumpa J, Bazán Alfaro I (1989) Capitalismo y formación regional: Chiclayo entre los siglos XIX y XX. Población y Desarrollo Instituto de Investigación y Capacitación, Chiclayo, p 216
Gonzales de Olarte E, Samamé L (1991) El péndulo peruano: políticas económicas, gobernabilidad y subdesarrollo en el Perú 1963–1991. Lima: IEP, 129 p
Huerta L, Mateo D, Mesclier É (1997) Perú en mapas. Atlas en base al censo de población y vivienda. Lima: INEI-ORSTOM, 136 p
Huetz de Lemps (1983) "L'évolution récente de la culture de la canne à sucre au Pérou" dans Huetz de Lemps, A. et Collin Delavaud, A., 1983. La canne à sucre en Espagne, au Pérou et en Equateur. Collection de la maison des pays ibériques num. 11. PARIS: Fondation Nationale des Entreprises Publiques, pp 43–95
INEI (1995) Migraciones internas en el Perú. Lima, INEI- FNUAP. https://www.inei.gob.pe/media/MenuRecursivo/publicaciones_digitales/Est/Lib0018/n00.htm
Klarén P (1970) Formación de las haciendas azucareras y orígenes del APRA. Lima: IEP, 298 p
Marshall A (2009) S'approprier le désert. Agriculture mondialisée et dynamiques socio-environnementales sur le piémont côtier du Pérou. Sous la direction de Jean-Louis Chaléard, Evelyne Mesclier et Frédéric Bertrand. Thèse de Doctorat. Université de Paris 1 Panthéon-Sorbonne. 493 p
Marshall A (2014a) Empresas agroindustriales, pequeña agricultura y acceso a la tierra en periferias de ciudades de la costa peruana. Panel: Democracia local y acceso de los agricultores a la tierra en las periferias urbanas. Conférence présentée au LASA2014 Democracia y Memoria, XXXII Congreso Internacional de la Asociación de Estudios Latinoamericanos. 21–24 de mayo del 2014. Chicago
Marshall A (2014b) Apropiarse del desierto. Agricultura globalizada y dinámicas socioambientales en la costa peruana. El caso de los oasis de Virú e Ica-Villacuri. IFEA-IRD, 417 p
Matos Mar J (2004) *Desborde popular* y crisis del Estado. *Veinte años* después. Lima, Fondo Editorial del Congreso de la República
Mesclier É (2000) Trente ans après la réforme agraire péruvienne: le discours des investisseurs contre la voie paysanne. Problèmes d'Amérique latine, juillet-septembre 2000, pp 117–141
Mesclier É (2009) Le titrage des terres, instrument d'une restructuration orientée de l'agriculture péruvienne, In: Colin J-P, Le Meur P-Y et Léonard E, Les politiques d'enregistrement des droits fonciers. Du cadre légal aux pratiques locales. Paris, Karthala, pp 445–475
Mesclier É, Marshall A, Chaléard J-L, Auquier C (2013) L'agriculture entreprenariale d'exportation: un choix politique aux enjeux complexes. Problèmes d'Amérique latine n° 88, Printemps 2013, Dossier Pérou: émergence économique et zones d'ombre (I. Hurtado, É. Mesclier, coord.), pp 55–76
Oré MT (2005) Agua. Bien común y usos privados. Riego, Estado y conflictos en la Achirana del inca. Lima, Pontifica Universidad Catolica del Peru, 246 p

Oré MT, Bayer D, Chiong J, Rendón E (2012) Emergencia hídrica y explotación del acuífero en un valle de la costa peruana: el caso de Ica. In Raúl H. Asensio, Fernando Eguren y Manuel Ruiz, eds. Péru: el problema agrario en debate. SEPIA XIV. Lima: SEPIA, 2012, 798 pp

PECH Proyecto Especial Chavimochic (2010) Chavimochic en cifras 2000–2010. Creciendo juntos en Libertad. 97 p; Trujillo. Gobierno regional La Libertad

Piel (1975) Capitalisme agraire au Pérou: originalité de la société agraire péruvienne au XIXe siècle. Paris: Anthropos, vol 1, 330 p

Piron M, Mesclier É, Lortic B (2015) Expansion de l'agglomération de Lima et différenciation de l'espace résidentiel: analyse exploratoire d'un corpus de données diversifié, Cybergeo: Eur J Geogr [online], Dossiers, Images & Villes, document 732, mis en ligne le 23 juin 2015

Pulgar Vidal J (1940) Las ocho regiones naturales del Perú. Lima, 16 p

Retaillé D (2006) Oasis. In: Levy J, Lussault M (eds) Dictionnaire de la géographie et de l'espace des sociétés. Belin, Paris, 1033 p

Revesz B (1991) Pérou: vingt ans après la réforme de la périphérie agraire, les impuissances de l'État. *Problèmes d'Amérique latine*, octobre-décembre 1991, pp 3–26

Reynaud A (1983) Centre et périphérie. In: Bailly A, Ferras R, Pumain D (eds) Encyclopédie de géographie. Paris, Economica, pp 583–599

Rostworoski de Diez Canseco M (1981) Recursos naturales renovables y pesca, siglo XVI y XVII, 180 p, Lima: IEP

Schlüpmann J (1994) La structure agraire et le développement d'une société régionale au nord du Pérou. Piura, 1588–1854. 2 tomes. Thèse de Doctorat. UFR Géographie, Histoire et Science de la Société. Université de Paris VII, Paris

Swyngedouw E (1997) Neither Global nor Local. "Glocalization" and the Politics of Scale. In: K. Cox Spaces of Globalization. Reasserting the power of the local. New York, London, pp 137–166

Vasquez F (2000) Actualización de los estudios del nivel freático en los valles de Chao, Virú y Moche. FAO proyecto TCP/PER/8823. Proyecto Especial Chavimochic. 45 p

Internet Sites

JURP Moche Virú Chao: http://www.jriegopresurizado.org.pe/
Camposol S.A.: http://www.camposol.com.pe/
Tesco superstore: http://www.tesco.com/nurture/?page=nurturestandards
Global Gap: http://www.globalgap.org/uk_en/

Part II
History of Water and Oases

The Genesis of Oases in Southeast Arabia: Rethinking Current Theories and Models

Julien Charbonnier

Abstract In South East Arabia (Sultanate of Oman and United Arab Emirates), oases are irrigated gardens characterised by intensive and mixed farming: date palms form a canopy under which other crops are protected from the sun's rays and the heat. The origin of this agrosystem and its impact on the historical trajectory of Arabian populations are still much debated issues. Some scholars have suggested that oases developed as soon as agriculture was introduced into the region, at the beginning of the Early Bronze Age (3200–2000 BC). The intensification of trade with neighbouring civilisations of Mesopotamia, Iran and Indus seems to have ignited the spark for the adoption of agriculture in Southeast Arabia. According to the existing theories, oases then emerged rapidly and were a means of adapting agriculture to the arid environment of this region. This agrosystem remained unchanged in its fundamental principles until the present day. This theory, however, denies the diachronic dimension of Southeast Arabian landscapes. The present chapter therefore aims to re-evaluate the data and reassess current theories of an Early Bronze Age origin for oases. An alternative development model, based on the available data, will be proposed. It is suggested that the development of oases corresponds to a long process with several steps resulting from environmental changes, technological innovation and socio-economic factors.

Keywords Oasis · Arabia · Protohistory · Archaeobotany · Palaeoclimatology · Hydraulic

> 'Is this Arabia,' we said, 'is this the country we have looked on heretofore as a desert?' Verdant fields of grain and sugarcane, stretching along miles, are before us; streams of water, flowing in all directions, intersect our path; and the happy and contented appearance of the peasants agreeably helps to fill up the smiling picture. The atmosphere was delightfully clear and pure. (James Wellsted's travel account cited in Hogarth 1904: 138–139).

J. Charbonnier (✉)
UMR 7041 ArScAn (Archéologies et Sciences de l'Antiquité), Maison René-Ginouvès, 21 allée de l'Université, 92023 Nanterre Cedex, France
e-mail: julcharbonnier@gmail.com

E. Lavie and A. Marshall (eds.), *Oases and Globalisation*,
Springer Geography, DOI 10.1007/978-3-319-50749-1_4

Fig. 1 Spatial organisation of an oasis (**b**) and vertical distribution of crops in the palm grove (**a**)

In Arabia and North Africa, the oases correspond to irrigated gardens characterised by intensive and mixed farming. The date palm is the keystone species of this agrosystem, also called "*bustān*" in the scientific literature. Palm trees form a canopy under which other crops are protected from the sun's rays and the heat in these arid regions (Fig. 1a). The temperature is always lower in the palm grove than in the surrounding areas. Evaporation from canals and the surface of flooded plots is also greatly reduced (Laureano 1998). Under the date palms, crops are distributed vertically; two to three layers can be distinguished: from top to bottom: date palms, fruit trees (lemon trees, fig trees, etc.), cereals and then vegetables and legumes (Tengberg 2012: 139–140). When a plot is irrigated, water feeds the palm trees, the other fruit trees and also annual plants. All the crops therefore benefit from permanent moisture in the soils. Wild plants and animals are also attracted by the oasis and contribute to the formation of soils. Fields are rarely left fallow and can provide one harvest of dates and other fruits and several harvests of cereals and vegetables annually. Apart from the palm grove, the other components of an oasis are the irrigation system, the open fields and the settlements (Fig. 1b). Cereals and forage crops can be grown in open fields, located downstream or around the palm grove. Hydraulic structures extract water from the ground or divert surface run-offs in order to feed the oasis and keep it green all year long. The last element, a key component of an oasis, is intangible: it consists of the rules and the organisation adopted by a community of people to share water and maintain the irrigation system.

Established in the piedmont area, on coastal plains, or terraced on mountain slopes, oases are landmarks of Southeast Arabia (i.e. the Sultanate of Oman and the United Arab Emirates). Located between the Persian Gulf and the Gulf of Oman, this area is characterised by an arid climate. However, moderate rainfall occurs during winter on the Hajar Mountains, which separate the coastal plain from the inland desert (Sanlaville 2000: 49). Large floods, which form after rainfall events, dry out on the piedmonts surrounding the mountains and recharge the underflows.

The region also benefits from fossil aquifers. Water resources are traditionally exploited with the help of hydraulic devices locally called *aflāj* (sing. *falaj*). This term actually refers to several types of structures. On the piedmont of the Hajar Mountains, most *aflāj* correspond to *qanāts*, tunnels draining groundwater to the surface, while in the mountain valleys, *aflāj 'ayni* usually tap springs. The term *falaj ghayli* is used for systems diverting floodwaters.

The study of water management techniques and oases is crucial to understand the historical trajectory of local societies in the second half of the Holocene as irrigation was essential to sustain agriculture. The development of new agrosystems and new hydraulic techniques may have had a deep impact on the settlement pattern, population growth and trends towards specialisation or the increase in hierarchy. Some scholars have suggested that they developed as soon as agriculture was introduced into the region, at the beginning of the Early Bronze Age (3rd millennium BC).

The present chapter is devoted to the question of the development of this agrosystem in Southeast Arabia and aims to re-evaluate the data and reassess current theories of an Early Bronze Age origin for oases. An alternative development model, based on the available data, will be proposed. The main objective of this paper is thus to show that there is still an open debate on this question. Finally, it will be argued that further progress on this issue will depend on multidisciplinary approaches and the excavation of palm groves.

1 The "3rd Millennium BC Oases" Theory

During the Umm an-Nar period, Southeast Arabian societies underwent significant changes. Sites were integrated into a regional exchange network with the export of raw materials—mainly copper—to Mesopotamia and the Indus valley (Cleuziou 1999: 99). Ceramic technology and monumental stone and mud brick architecture appear for the first time in this region (Cleuziou and Tosi 2007: 128). Collective burials, built above ground, could reach several metres in diameter and were sometimes made of standing stones. Massive circular or subcircular "towers", sometimes surrounded by ditches, were built on the main sites. Their function is still debated, but they required a considerable investment. At that time, society was characterised by social stratification with a concentration of wealth in the hands of the few although it seems that many decisions might have been taken collectively. In addition, this period witnessed demographic growth, hence the need for agriculture to sustain the population (Cleuziou 2005: 144).

The most widely accepted theory concerning the emergence of oases in Southeast Arabia was developed by Serge Cleuziou, who was responsible for the study of the Early Bronze Age (3200–2000 BC) site of Hīlī 8, in the oasis of al-'Ayn, UAE (Fig. 2). During the excavation of Hīlī, numerous remains of wheat, barley and date stones have been found bearing testimony to the practice of agriculture since the first half of the 3rd millennium BC (Costantini 1979; Cleuziou and

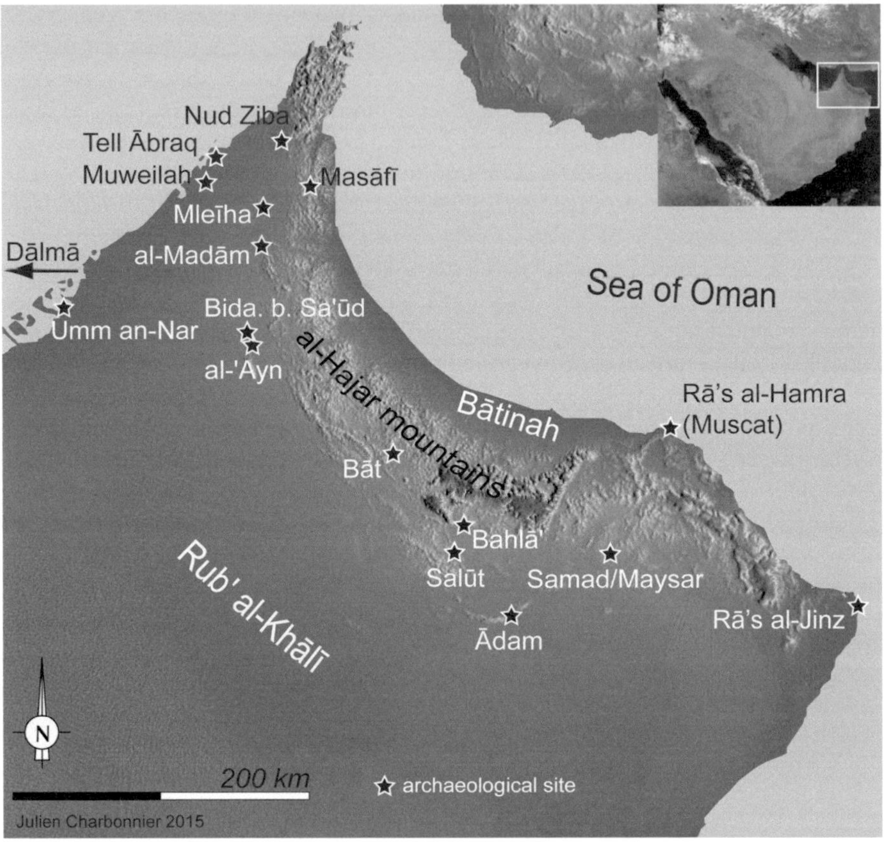

Fig. 2 Map of Southeast Arabia with sites mentioned in the text

Costantini 1980). Later, Cleuziou made the assumption, based on the archaeobotanical discoveries in Hīlī and other contemporary sites, that the oasis agrosystem already existed during that period: the cereals would have been cultivated under a date palm canopy (Cleuziou 1982: 19–20, 1997: 391–392, 2009: 730). Based on the identification of the charred remains and imprints of sorghum from Hīlī 8, Cleuziou also suggested that annual crops were cultivated throughout the year (i.e. wheat and barley as winter cereals harvested in spring and sorghum as a summer crop harvested in autumn) as in present-day oases (Cleuziou 1997: 401). This idea was taken up by other researchers such as Potts (1999: 36), Méry (2013) and Magee (2014: 2390). Margareta Tengberg, an archaeobotanist, also suggested that the oasis agrosystem was already underway in the 3rd millennium BC owing to the simultaneous presence of cereal, vegetable, legume and date stone remains on Bronze and Iron Age sites (Tengberg 2003; 2012).

Cleuziou later suggested that the first stage in oasis development was the establishment of "small palm tree gardens" during the first half of the 3rd millennium BC, based on Jessica Giraud's results of a survey in the Ja'alan region, east of Oman (Cleuziou 2009: 734; Giraud 2007: 191). Concerning the Hafit period (3200–2700 BC), Giraud noticed that the cairn graves, which are highly visible in the landscape because of their prominent positions, were often located close to areas near present-day palm gardens. She suggested that Hafit graves were established in the past near small oasis settlements (Giraud 2009: 747–748).

Contacts with Mesopotamia, Iran and Indus would have ignited the spark for the adoption of agriculture as the cereals cultivated might have been introduced from these regions (Tengberg 2003: 235). Cleuziou thinks that the oasis is an indigenous and original response to these stimuli, i.e. that this agrosystem was invented locally in order to adapt agriculture to the arid environment of Eastern Arabia.

In Cleuziou's view, this agrosystem then remained unchanged until the present day. Therefore, the agricultural landscape has never really changed: "By analogy with the present-day oases of the Oman Peninsula, we may imagine that some plants were grown in the shades of these palm trees" (Cleuziou 1982: 19). Walid al-Tikriti shares the same vision when he assumes that "the landscape of the oases cannot have been very different from what it was before the recent oil-boom era" (al-Tikriti 2002: 137). For al-Tikriti, however, the oases developed in the first millennium BC, in parallel with the development of *qanāt* technology (al-Tikriti 2010: 240–243). Cleuziou, in contrast, considers that *qanāts* could have been already used in the 3rd millennium BC and were sustaining the oases from the beginning (Cleuziou 2009: 731).

Clearly, Cleuziou does not claim that oases have no historical trajectory. On the contrary, he is aware that the evolution of climate and water resources could have led to the abandonment or the spatial evolution of some oases. For instance, the lowering of the water table has led to the lowering of the *qanāts* and gardens in order to maintain gravity-fed irrigation, as observed during the twentieth century AD (Bisson 1989: 183; Costa 1983: 248–249; Weisgerber 2005: 75–78). However, he implies that the agrosystem remained the same because it was adapted to the arid climate of Arabia from the beginning.

Nevertheless, the theory of Cleuziou can be discussed and criticised in the light of climatic, archaeobotanical, sedimentary and archaeological data. Were the landscapes of the Bronze and Iron ages really similar to the present-day one? When and how did palm groves develop?

2 Climate Change During the Second Half of the Holocene in Arabia

The Holocene moist phase, lasting until the 5th millennium BC, was characterised by wetter climatic conditions (Fleitmann and Matter 2009: 640) resulting in the development of lakes and a savannah-like landscape in Arabia (Parker et al. 2004, 2006). There were more wild animals, and grazing was facilitated. During the so-called Neolithic period in Arabia, populations were therefore mainly hunter-pastoralist. Along the coasts, groups specialised in fishing, gathering shells and exploiting mangroves (Biagi and Nisbet 2006; Cleuziou 2005: 134). With the retreat of the monsoon towards the south, summer rains ceased to reach Southeast Arabia from the 5th millennium BC (Fleitmann and Matter 2009: 640); peaks of aridity then occurred at the beginning and end of the following millennium. The 4th millennium BC corresponded to a slightly moister period that was interrupted by an arid event around 3200–3000 BC. The climate became slightly wetter again, but remained arid during the 2nd millennium BC (Parker et al. 2006: 472–473). Aridity increased around 1000 BC: during the Iron Age, the climate was as arid as it is today (Fleitmann et al. 2007: 180).

Relatively little is known about the evolution of groundwater in Southeast Arabia during the Bronze and Iron ages. During the Umm an-Nar and Wadi Suq periods, it is clear that the water table retreated in al-ʻAyn from two wells that were excavated at Hīlī 8. The first one, dating from the first half of the 3rd millennium BC, was about 4 m deep, while the second one, used from *c.* 2700 to *c.* 1800 BC, reached 8.50 m in depth (Cleuziou 1989: 64–68). This suggests that the groundwater was quite close to the surface at the beginning of the Umm an-Nar period and that it dropped around 4.5 m over 1000 years. Similar processes seem to have taken place during the 1st millennium BC, as some *qanāts* were abandoned during or at the end of the Iron Age. The *qanāt* AM-2 excavated in al-Madām was probably abandoned after the local population desperately tried to lower the tunnel and the channels (Córdoba 2013: 148; Córdoba and Del Cerro 2005: 525).

3 Crops Cultivated During the Bronze and Iron Ages in Southeast Arabia

According to the recent studies, wild date palms (*Phoenix dactylifera*) were growing along the Persian Gulf and could have been domesticated locally (Gros-Balthazard et al. 2013: 17–19 and 24). The oldest remains of date stones found in this region come from the island of Dālmā (Abu Dhabi emirate, UAE) and have been radiocarbon-dated to the end of the 6th/beginning of the 5th millennium BC. They were recovered during the excavation of the Neolithic site Dālmā 11. The archaeobotanists who studied these two charred stones were unable to tell whether they corresponded to wild or cultivated dates (Tengberg 2012: 142). They suggested they could have been gathered in the region or imported from Mesopotamia (Beech and Shepherd 2001: 87–88). If reference is made to the above-mentioned study, it seems to me likely that these dates grew locally, maybe not on Dālmā Island but somewhere in Southeast Arabia. As regards the Neolithic period, data concerning the exploitation of date palms are therefore limited.

When it comes to the 3rd millennium BC, there are far more charred remains and evidence for the cultivation of date palms. At Hīlī 8, plant remains have been found in the oldest layers, dated by Cleuziou to the Hafit period, and hundreds of charred date stones in the Umm an-Nar levels (Cleuziou and Costantini 1980). Numerous charred stones have also been recovered in 3rd millennium layers in Rā's al-Jinz 2 (RJ-2, Oman) and Tell Ābraq (UAE, Costantini and Audisio 2001; Willcox and Tengberg 1995). There is also evidence of dates on the sites of Bāt (Oman, charred remains and impressions on mud bricks), Maysar 1 (Oman, charred remains) and Umm an-Nar (UAE, impressions on mud bricks) (Tengberg 1998: 188f; Weisgerber 1981: 191–197; Willcox 1995: 257–259). Margareta Tengberg notes that stem tissue remains of date palm have also been identified in Hīlī, which seems to indicate that this species was cultivated locally (Tengberg 2003: 232). On the whole, it is very likely that given the great quantity of date stones discovered and their presence in many excavated Umm an-Nar settlements, even coastal ones such as Tell Ābraq and RJ-2, date palms were already cultivated at that time in Southeast Arabia, especially since this species seems to have been endemic to the Gulf. Date palms probably grew next to inland sites, such as Bāt and Hīlī, and may have been brought to coastal sites in exchange for some goods (Cleuziou 1999: 98–99).

In comparison, there is little information about the Wadi Suq period (Middle Bronze Age, 2000–1600 BC). Date stones were discovered in Nud Ziba (UAE, Kennet and Velde 1995: 85). This fruit was also part of the diet, along with cereals, in Tell Ābraq (Tengberg 2003: 233). More recently, jujube and date stones were collected during the excavation of a Late 2nd millennium site (1400–1200 BC) at Masāfī, in the UAE (Degli Esposti and Benoist 2015: 65).

The range of crops attested during the Umm an-Nar period is quite limited. Bread wheat (*Triticum* cf. *aestivum*) was present in Bāt, Umm an-Nar, Tell Ābraq and Hīlī 8. In the latter site, there were also charred remains of emmer (*Triticum dicoccum*), oat (*Avena* sp.), two-rowed barley (*Hordeum vulgare* subsp. *distichum*), six-rowed barley (*Hordeum vulgar*e subsp. *hexastichum*) and peas (*Pisum sativum*) (Tengberg 2003: 232). Impressions of six-rowed barley were found in Bāt and Umm an-Nar. Charred remains and impressions from seeds of sorghum (*Sorghum bicolor*) were initially reported in Hīlī 8 by Costantini (Cleuziou and Costantini 1980), but this identification was seriously questioned later (Rowley-Conwy et al. 1999, Tengberg 2012: 145).

There is little information concerning the consumption of fruits. Charred seeds of melon (*Cucumis melo*) were found in Hīlī 8, while many charred remains of jujube (*Ziziphus spina-christi*) were discovered during the excavations of Hīlī 8, RJ-2 and Bāt (for an overview, see Costantini and Audisio 2001: Table 6). Jujube tends to be collected from the wild, so the trees were not necessarily cultivated. It was already being consumed during the 5th and 4th millennia BC, as demonstrated by finds in the Neolithic sites of Rā's al-Hamra 5 and 6 (Oman, Biagi and Nisbet 1992).

The agrosystem of the Bronze Age was therefore limited compared to that of the present day. Only winter cereals and pulses, originating from the Near East, seem to have been cultivated during the Umm an-Nar period (Tengberg 2003: 232). Not many fruit trees were exploited at that time: only date palms were cultivated with any certainty.

In the Iron Age, dates were an important part of the diet. Thousand five hundred charred stones were recovered at Muweilah (UAE), along with more than 1300 carbonised dates (Tengberg 1998). Analyses carried out on teeth from pre-Islamic graves in al-Maysar (Oman) showed that caries were rare in the Early Bronze Age and became more common during the Iron Age, which suggest that the proportion of dates in the diet increased during the 1st millennium BC (Kunter 1981; Potts 1990: 127–132). Impressions of free-threshing wheat (*Triticum aestivum/durum*) and barley (*Hordeum vulgare*) were identified in Rumeilah (al-'Ayn, UAE) and Tell Ābraq (Costantini and Costantini-Biasini 1986: 357–358; Willcox and Tengberg 1995: 133). A recent palynological study led in Salūt (Oman) has identified the presence of sesame (*Sesamum* sp.) in the Iron Age layers. The pollen amount in some levels could indicate the use of this plant for oil or flour production (Bellini et al. 2011: 2785).

During the Late pre-Islamic period (PIR, third century BC–third century AD), excavations at Mleīha (UAE) reveal that dates, winter cereals (barley, *Hordeum vulgare*, and wheat, *Triticum durum* or *aestivum*) and pulses (lentils, *Lens culinaris*, broad beans, *Vicia faba*, and grass peas, *Lathyrus sativus*) were the basis of the diet before the advent of Islam (Pena-Chocarro and Barron Lopez 1999: 64–68; Mouton et al. 2012: 214). Caries seem to have been common at that time, probably due to the high consumption of dates (Nelson et al. 1999). Fleshy parts of grapes (*Vitis vinifera*) and pomegranates (*Punica granatum*) were recently found in a building dated between the second and the mid-third century AD at Mleīha (Mouton et al. 2012: 214). They could have been cultivated locally, but this has still to be confirmed.

4 Soil Studies of Oases

There is a clear lack of pedosedimentary data concerning oases in Southeast Arabia. Soil studies of palm groves are almost non-existent. A recent micromorphological analysis of a naturally exposed crosscut in *wādī* al-Sharsā, near Bāt, suggests that floodwaters were used from the 4th millennium BC in this area. Before that date, the *wādī* was an active floodplain, which then became less active. A subsequent gradual silting of the plain, interspaced with episodes of rapid flooding, testifies to anthropic flood diversion, i.e. fields were irrigated with *wādī* flows. Irrigation seems to have started in the second half of the 4th millennium BC and then became more intensive with evidence of tillage in the 3rd millennium BC (Desruelles et al. 2016).

5 The Pre-Islamic Hydraulic Techniques in Arabia

All oases depend on irrigation systems that collect run-offs or floods or tap groundwater. As already mentioned, a recent study in Bāt seems to indicate that floods were already diverted towards the fields during the 4th millennium BC in *wādī* al-Sharsā. Although dikes and dams are used today, only one can be dated to the Bronze Age with certainty. It is a 300-m-long low wall that closes a small valley in the region of Bahlā' (Oman). It could have been used to retain water and sediments for agriculture or grazing. Several ingots and anthropomorphic figures made of bronze have been recovered from its masonry. The plano-convex ingots are generally dated to between the 3rd and 2nd millennia BC. The figures are similar to some others found in occupation layers from the same period at Rā's al-Jinz (Weisgerber and Yule 2003: 48–51).

Two possible dams were also reported by a German mission in wādī Samad (Oman), where many Bronze and Iron Age sites have been identified, some of which have been excavated. These two small levees of stone blocks are located next to a *wādī* channel and near an Umm an-Nar settlement, which is why they have been interpreted as Bronze Age diversion dams. It has been suggested that they were used to trap water and sediments (Hastings et al. 1975: 11). These structures are therefore poorly dated. The same remark applies to a series of walls near Bāt, located at the bottom of a slope and perpendicular to a *wādī* bed. Unfortunately, no OSL or radiocarbon analysis has been carried out in the surrounding layers. An Umm an-Nar sherd, recovered at the bottom of one of the walls and partly slid up under it (Brunswig 1989: 22–25), only provides a *terminus post quem*. The function of these structures is unknown, but it seems they are unlikely to correspond to diversion dams as Brunswig and Frifelt suggested (Frifelt 1985: 99). Indeed, unlike most diversion dams, they are not slanted compared to the *wādī* bed and they are too close to each other.

In the recent past, groundwater was extracted through wells and *qanāts* (Fig. 3). The oldest recorded wells in Southeast Arabia date back to the 3rd millennium BC. As already mentioned, two of them have been excavated at Hīlī 8. They were located inside Umm an-Nar buildings and are accurately dated, but they were probably not used for irrigation (Cleuziou 1989: 64–68). Wells have been discovered in many Early Bronze Age settlements such as Bāt (Frifelt 2002: 104), Salūt (Degli Esposti 2011) and *wādī* Samad (Weisgerber 1981: 203). During the Iron Age, they were used to provide water to cattle and humans (al-Madām-1, UAE, Córdoba and Del Cerro 2005: 519–520), for manufacturing mud bricks (al-Madām workshop, Córdoba 2013) and for irrigation. Two wells associated with channels

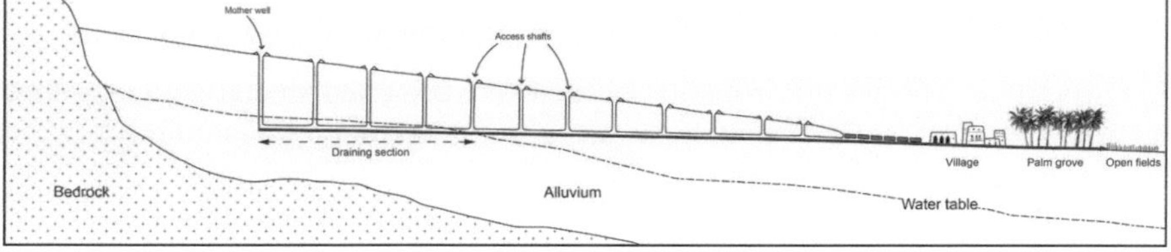

Fig. 3 Section of a *qanāt*

were found during a rescue excavation in Qattārah (al-ʻAyn, UAE), while another was associated with tree pits. These structures come from a stratified context and could be dated with certainty to the Iron Age II (Power and Sheehan 2011: 272–273). One well was excavated in Mleīha in area B, which could be dated to the Late pre-Islamic period (Dalongeville 1999: 43). Michel Mouton thinks that wells were used in this plain to irrigate garden crops and date palms (Mouton 1999: 272).

The development or introduction of *qanāt* technology in Southeast Arabia is still debated. Some scholars have suggested that *qanāts* were already in use during the Early Bronze Age. A partly excavated *qanāt* in the region of Bahlā' has been dated to the 3rd millennium BC (Orchard and Orchard 2007: 150–151), but the arguments of the excavators have proved insufficient (see Charbonnier 2015). Cleuziou has also argued for the use of *qanāts* at Hīlī 8 where some ditches and channels were lowered several times. He suggested this was related to the lowering of the *qanāt* tunnel in response to the retreat of the water table during the 3rd millennium BC (Cleuziou 1998: 61–62, 1999: 90, 2009: 731). However, no remains of Bronze Age *qanāts* have been found so that Cleuziou's idea remains an assumption.

It is now agreed among scholars working in Southeast Arabia that *qanāts* were used from the beginning of the Iron Age II. Ancient *qanāts* have been excavated at Hīlī, Bida Bint' Saʻūd and al-Mādam (al-Tikriti 2002 and 2010; Cordoba and Del Cerro 2005; Cordoba 2013). Others have been reported after surveys (al-Tikriti 2002; Schreiber 2007: 136; Weisgerber 1981). I have argued elsewhere that only the al-Mādam *qanāt* (AM-2) was well dated to the Iron Age, thanks to the excavation of the tunnel and the associated surface channels and fields. A radiocarbon date around the beginning of Iron Age II was also obtained on a gastropod shell that appears to have lived in the channels (Córdoba 2013: 147–148). There are still some doubts concerning Hīlī and Bida Bint' Saʻūd, because their dating relies mainly on their proximity to Iron Age settlements and the discovery of pottery sherds during their excavation, whereas there are very few elements that confirm the age of the surveyed structures (Charbonnier 2015).

Data concerning pre-Islamic hydraulic structures are still fragmentary, as there is no information regarding the 2nd millennium BC. Floods and run-offs could have been used from the end of the 4th or the 3rd millennium BC. Wells were dug during the Umm an-Nar period and throughout the Iron Age. *Qanāts* were probably not introduced before the 1st millennium BC.

6 Water Management in Southeast Arabia

In arid regions, water from irrigation systems is very often shared among a community. There are no data about the pre-Islamic hydraulic communities, as the Bronze and Iron ages are prehistoric in nature, so we have to look at water management during modern times to try to understand the impact of hydraulic structures on societies and landscapes.

From a management perspective, there are broadly two categories of hydraulic structures: those that can be built at an individual or familial level and those that have to be built by a community or political entity. Although wells can be beneficial to a community, they fall into the first category (Wilkinson 1977: 97). A family can dig them, especially if the water table is located a few metres below the surface as was the case during the Early Bronze Age in some areas (cf. Hīlī 8). Garden plots tend to be clustered around each well and are fed by a radial network of channels extending outwards, as was the case in the Bāṭinah region (Costa and Wilkinson 1987: 43–53). Plots and dwellings are not necessarily grouped but can be scattered (Wilkinson 1977: 97).

Digging a *qanāt* or diverting floods requires much more labour investment (Beaumont 1989: 27; Costa and Wilkinson 1987: 37). In Southeast Arabia, *qanāts* belong to communities of people. The digging and maintenance (cleaning underground galleries and channels) are funded by all shareholders. One person—*wakīl*—is generally responsible for supervising these tasks, coordinating the work and collecting the money. This person is generally a prominent member of the oasis population, as he also settles the quarrels concerning the access to water. The water distribution can be managed by the shareholders or by a supervisor (*'arīf*). In Oman, UAE, Saudi Arabia and eastern Yemen, water shares from *qanāts* generally correspond to a period of time that is distributed along a water cycle. Indeed, the flow cannot be interrupted and must be shared day and night. In Oman, during the daytime, sundials were traditionally used to measure water shares, while stargazing was used at night (Al-Ghafri et al. 2004; Charbonnier 2014; Dutton 1989; Nash 2011).

The use of *qanāts* entails grouping the gardens as water must reach all of them and moves constantly from one field to another inside the palm grove (Charbonnier 2014: 92). To reduce the travelling time for water, fields must be as close as possible to each other. Thus, it seems important to group the fields as shareholders must be able to check whether the shares are being respected. In Ādam, people would come regularly to the sundial to see whether everything was all right concerning the water distribution. Numerous quarrels still arise showing the importance of discussion and transparency in the day-to-day management of community irrigation systems.

7 Discussion: Challenging Previous Views Concerning the Emergence of Oases

During the discussion, I will try to demonstrate three things: (1) agriculture emerged during the Bronze Age in Southeast Arabia, (2) Cleuziou's oasis theory is not firmly established and (3) the available data can be used to propose another model for the development of the oasis agrosystem in this region.

7.1 Bronze Age Agriculture in Southeast Arabia

Domesticated crops (wheat, barley, etc.) imported from the Near East have been present since the 3rd millennium BC in Southeast Arabia. At that time, local societies had closer contacts with Mesopotamia, Iran and the Indus valley, which explains their presence. As we have seen, archaeological evidence suggests that at least part of the population had adopted a sedentary lifestyle and society had become more stratified since at least the Umm an-Nar period. Archaeobotanical evidence suggests that annual crops and date palms were cultivated in Southeast Arabia at that time, i.e. cereals and dates were not imported from abroad. Indeed, cereal husks were also present on the sites as they were used as temper for mud bricks, which suggests that grains were threshed locally (Willcox and Tengberg 1995).

Al-Tikriti has suggested that *Phoenix* was not cultivated before the Iron Age in the region; dates would have been imported from abroad (al-Tikriti 2010: 243, footnote 23). In my view, this is unlikely, as large quantities of date stones have sometimes been found. Besides, archaeobotanists have demonstrated that the presence of stem tissue points to the local cultivation of date palms (Tengberg 2003: 232). This idea would also be strengthened by an indigenous origin of the date palm (Gros-Balthazard et al. 2013).

7.2 The 3rd Millennium BC Oases: A Questionable Hypothesis

Although agriculture seems to have been established during the 3rd millennium BC, the same does not apply to the "multi-storied" oasis agrosystem. Only archaeobotanical evidence has been used to demonstrate its existence, which, in my view, is not sufficient. As already noted by al-Tikriti, the discovery of cereals and date stones in an archaeological context does not prove that the former were cultivated below the latter (al-Tikriti 2010: 243, note 23). It only demonstrates that these crops were cultivated at the same time. Date palms and annual crops could also have been cultivated in separate fields, adjoining or spatially separated.

I will go even further by saying that nothing in the archaeobotanical data shows that the vertical distribution of plants typical of the oases existed at that time. On the contrary, the range of crops cultivated in the 3rd millennium BC and up to the Late pre-Islamic period was limited: it only included cereals, some legumes and date palms. Cultivated fruit trees are absent from the documentation until the Late pre-Islamic period, with the exception of jujube trees that do not require the protection of date palms to grow. This proves that, at best, the hypothetical pre-Islamic oases were composed of only two layers and not three or four as in the present day.

Besides, the annual plants attested during the Bronze and Iron ages are all winter (wheat, barley, oats and peas) or spring (oats) crops, which means that they could not be cultivated all year long as in present-day oases. Unless our data are incomplete, this means that the agrosystem of the Bronze and Iron ages was not as intensive as it is today. Furthermore, winter cereals do not have to be cultivated under date palms as they are grown during the coldest season, and in fact, they are generally cultivated in the open fields surrounding the palm groves. This allows me to hypothesise that date palms and annual crops were spatially separated until at least the Iron Age.

7.3 A Dynamic Landscape and the Gradual Formation of the Oases: An Alternative Model

Based on the available data, I propose an alternative, albeit hypothetical, model for the development of the oases in Southeast Arabia. I suggest that oases formed gradually following many changes in the landscape and the social structure. These changes were also enabled by the development or introduction of new hydraulic techniques.

(1) At the end of the Neolithic, populations were already taking advantage of wild plants (Tengberg 2003: 232). The date stones found at Dālmā are likely to correspond to the exploitation of wild date palms, as it is now highly probable that these were growing along the shores of the Persian Gulf.

(2) From the second half of the 4th to the beginning of the 3rd millennium BC, data from Bāt suggest that floodwater was already diverted at the foot of the Hajar Mountains. This still has to be confirmed, but as Fouache et al. (2012) have noted, it would show that the development of agriculture in this region predated the development of contacts with Mesopotamia and Iran in the 3rd millennium BC. However, the crops cultivated at that time have not yet been recognised. The floods could have been used only to cultivate winter crops, as there was already a winter rainfall regime. An interesting outcome of floodwater irrigation would have been the development of arable land. In arid regions, soils are sometimes as rare as water and *qanāts*, unlike surface water irrigation systems, do not provide much sediment to the fields. The areas where sediments brought by the flood irrigation accumulated could have been reused to install oases fed by *qanāts* in some places.

    Agricultural practices are therefore not confirmed during the Hafit period, but if data from Bāt are correct, they are likely to have occurred in some valleys of Southeast Arabia. It has been suggested that Hafit graves, located in prominent positions, marked the territory of nomadic pastoral populations when they were absent (Deadman 2012: 33). It is likely that many groups in the 3rd and even in the 4th millennium BC maintained this lifestyle, but this does not exclude the possibility of the existence of agricultural communities in some areas, for example the water-rich mountains, or of semi-nomadic groups practicing irrigation or rainfed agriculture in addition to pastoral activities. In parallel, it is known that coastal populations had a halieutic economy. Date palm domestication has yet to be understood, but these trees could have been exploited since that time. Battesti (2005) mentions the existence of semi-nomadic groups in Sahara who own palm gardens but do not irrigate them. The trees take advantage of groundwater relatively close to the surface. Date palm roots seem to be able to reach water up to 17 m below ground (Peyron 2000: 12). In fact, feral palms can grow on *wādī* banks or near springs. Only a few members of the group stay next to the gardens all year long. They only carry out pollination, harvest and vegetative propagation (by planting suckers). Such practices are attested in Chad and Sudan (Battesti 2005: 24). These non-irrigated palms are less productive but require less labour investment. Semi-nomadic groups, at the end of the Neolithic and at the beginning of the Bronze Age, might well have exploited such palm gardens.

(3) During the Umm an-Nar period, agriculture seems established but not the multi-storied oasis agrosystem. No Bronze Age irrigation system in relation to date palm plantations has yet been found in Southeast Arabia. I suggest that in some areas, such as Hīlī, date palms could have grown without a water supply due to the proximity of the water table. Their roots could have easily reached the shallow groundwater, especially along *wādī* banks. Wells are attested and could also have been used to water the palms, as well as other crops, in other areas. Cereals and legumes had to be irrigated at that time because of the scarcity of rainfall. They would have been cultivated outside the palm grove, in open fields, during the winter season.
As already noted, individuals or families can own wells as the labour investment to dig them is generally limited. Consequently, water drawn from them does not have to be shared among the community and gardens do not have to be clustered: they are distributed in space along the aquifer (Costa and Wilkinson 1987: 43–53). The Bronze Age of Eastern Arabia was perhaps characterised by such agricultural plots spread across the landscape.
Unfortunately, there are no data from the arid event at the end of the 3rd millennium BC to the end of the 2nd millennium BC, and it is not known whether agricultural practices evolved during this period.

(4) The Iron Age saw an increase in arid conditions and, possibly, the development of a hydraulic technique enabling the exploitation of groundwater: the *qanāt*. Although the dating of several *qanāts* from the Iron Age is still debated (Charbonnier 2015), this technology seems attested in some places. In parallel, it is very likely that wells were still used during the 1st millennium BC. Alternatively, other scholars have suggested that *qanāts* were introduced only during the Islamic period (Power and Sheehan 2012: 303). In any case, the main point I wish to emphasise is that the introduction or development of *qanāt* technology must have had a profound impact on both the social organisation and the agrosystem:

– It led to the formation of cohesive groups bound by the management of a common resource. The water of a *qanāt* must be shared permanently, day

and night, and year-long. It also needs constant maintenance, which means that more members of the group had to adopt a sedentary lifestyle.
- Fields had to be clustered to facilitate water transport and the control of sharing.

I wonder whether the need to cluster gardens and fields, in order to share the water of *qanāts*, was one of the main factors leading to the formation of palm groves. In order to maximise the use of this shared water, populations would have started to cultivate crops under the date palms. The oasis agrosystem would thus have been an outcome of this process. It offered benefits in an environment that became even more arid after 1000 BC.
However, it is not certain when summer crops were imported into Southeast Arabia; archaeobotanical data suggest that only winter crops were cultivated at that time. The introduction of summer crops might have been facilitated by palm grove agriculture as it offered shade during the hot Arabian summers.

(5) Throughout the Late pre-Islamic and Islamic periods, the oasis agrosystem was enriched by the gradual importation of new cultivars (summer crops and fruit trees).

8 Conclusion

As Tim Power and Peter Sheehan have noted, hypothetical Bronze and Iron Age oases are often "retrospective projections of the present date-palm oasis onto the past" in the minds of Western scholars (Power and Sheehan 2012: 296; see also Giraud 2007: 190). More broadly, this approach to Arabian oases fits into the "timeless" vision of Arabia, its inhabitants and its landscapes, which is common to many Westerners from the first travellers to contemporary scholars (Magee 2014: 367 and 463).

The theory proposed in this article, although hypothetical, tries to consider the oasis-*bustān* agrosystem as the product of a long history and several feedback processes, from the domestication of date palms (local?) to the appearance of multi-storied crops related to the need to group gardens to share the water of *qanāts*. Clearly, a large variety of sources must be taken into account to understand the history of the oases and the evolution of landscape in South Arabia. Until recently, studies have mainly focused on archaeobotanical data and the excavation of hydraulic structures. The context of the latter is generally not well understood. A multidisciplinary approach is needed to fill this gap and should include soil and environmental studies. A recent project led by Louise Purdue and the author in Masāfī (UAE) is heading in this direction: test pits have been dug in the palm grove in order to reconstruct its evolution, and analyses (in micromorphology, pedology, archaeobotany, etc.) are on-going (Charbonnier et al. in press).

This appears fundamental to comprehend the genesis of oases as this agrosystem, which improves land and water use, most certainly had a profound impact on the settlement dynamics in arid regions of Africa and Arabia. It will also raise the question of the adaptation of human populations to the aridification process in the second half of the Holocene.

Acknowledgements The author would like to acknowledge the Fyssen Foundation for its support through the granting of a postdoctoral fellowship, which enabled him to carry out a study of Southeast Arabia oases. The present article is an outcome of this work. He is also very grateful to the late Prof. Tony Wilkinson (Durham University) for its kind and wise advice.

References

Al-Ghafri A, Inoue T, Nagasawa T (2004) Irrigation Scheduling of Aflaj of Oman: Methods and Modernization. University of Hokkaido. http://inweh.unu.edu/wp-content/uploads/2013/05/Traditional-Water-Distribution-in-Aflaj-Irrigation-Systems-of-Oman.pdf. Accessed 8 Feb 2015

Al-Tikriti WY (2002) The south-east Arabian origin of the *falaj* system. Proc Semin Arab Stud 32:117–138

Al-Tikriti WY (2010) Heading north: an ancient caravan route and the impact of the *falaj* system on the Iron Age culture. In: Avanzini A (ed) Eastern Arabia in the First Millennium B.C., Arabia Antiqua 6, "L'Erma" di Bretschneider, Rome, pp 227–247

Battesti V (2005) Jardins au désert, Évolution des pratiques et savoirs oasiens, Jérid tunisien. IRD éditions, Paris

Beaumont P (1989) The qanat: a means of water provision from groundwater sources. In: Beaumont P, Bonine M, McLachlan K (eds) Qanat, Kariz and Khattara: traditional water systems in the Middle East and North Africa. School of Oriental and African Studies, University of London and Middle East and North African Studies Press, London and Cambridgeshire, pp 13–31

Beech M, Shepherd E (2001) Archaeobotanical evidence for early date consumption on Dalma Island, United Arab Emirates. Antiquity 75(287):83–89

Bellini C, Condoluci C, Giachi G, Gonnelli T, Lippi MM (2011) Interpretative scenarios emerging from plant micro- and macroremains in the Iron Age site of Salut, Sultanate of Oman. J Archaeol Sci 38(10):2775–2789. doi:10.1016/j.jas.2011.06.021

Biagi P, Nisbet R (1992) Environmental history and plant exploitation at the aceramic sites of RH5 and RH6 near the mangrove swamp of Qurm (Muscat-Oman). Bulletin de la Société Botanique Française, Actualités botaniques 139: 571–578

Biagi P, Nisbet R (2006) The prehistoric fisher-gatherers of the western coast of the Arabian Sea: a case of seasonal sedentarization? World Archaeol 38(2):220–238. doi:10.1080/00438240600708188

Bisson J (1989) The origin and evolution of foggara oases in the Algerian Sahara. In: Beaumont P, Bonine M, McLachlan K (eds) Qanat, Kariz and Khattara: Traditional Water Systems in the Middle East and North Africa. School of Oriental and African Studies, University of London and Middle East and North African Studies Press, London and Cambridgeshire, pp 179–194

Brunswig RH (1989) Cultural History, Environment and Economy as seen from an Umm an-Nar Settlement: Evidence from Test Excavations at Bāt, Oman, 1977/78. J Oman Stud 10:9–50

Charbonnier J (2014) In the shadow of the palm trees: time management and water allocation in the oasis of Ādam (Sultanate of Oman). Proc Semin Arab Stud 44:83–98

Charbonnier J (2015) Groundwater management in Southeast Arabia from the Bronze Age to the Iron Age: a critical reassessment. Water Hist 7(1):39–71. doi:10.1007/s12685-014-0110-x

Charbonnier J, Purdue LE, Calastrenc C, Régagnon E, Sagory T, Benoist A (in press) Ancient agricultural landscapes in Southeast Arabia: Approach and first results of an archaeological, geo-archaeological, and spatial study of the Masāfī Palm Grove, Emirate of Fujairah. In: al-Tikriti W (ed) Proceedings of the water and life in Arabia conference, Al-Ain

Cleuziou S (1982) Hili and the beginning of oasis life in Eastern Arabia. Proc Semin Arab Stud 12:15–22

Cleuziou S (1989) Excavation at Hili 8: a preliminary report on the 4th to 7th campaigns. Archaeology in the United Arab Emirates V: 61–87

Cleuziou S (1997) Construire et protéger son terroir: les oasis d'Oman à l'Âge du Bronze. In: Burnouf J, Bravard J-P, Chouquer G (eds) La dynamique des paysages protohistoriques, antiques, médiévaux et modernes. XVIIe Rencontres Internationales d'Archéologie et d'Histoire d'Antibes. Éditions APDCA, Antibes, pp 389–412

Cleuziou S (1998) The foundation of early bronze age Oases in the Oman Peninsula. In: Pearce M, Tosi M (eds) Papers from the EAA Third Annual Meeting at Ravenna 1997, vol I: Pre- and Protohistory. BAR International Series 717, Archaeopress, Oxford, pp 59–70

Cleuziou S (1999) Espace habité, espace utilisé, espace vécu dans les communautés d'Arabie orientale à l'Âge du Bronze. In: Braemer F, Cleuziou S, Coudart A (eds) Habitat et société. XIXe Rencontres Internationales d'Archéologie et d'Histoire d'Antibes. Éditions APDCA, Antibes, pp 83–108

Cleuziou S (2005) Pourquoi si tard? Nous avons pris un autre chemin. L'Arabie des chasseurs-cueilleurs de l'Holocène à l'Age du Bronze. In: Guilaine J (ed) Aux marges des grands foyers du Néolithique. Errance, Paris, pp 123–148

Cleuziou S (2009) Extracting wealth from a land of starvation by creating social complexity: a dialogue between archaeology and climate? CR Geosci 341(8–9):726–738. doi:10.1016/j.crte.2009.06.005

Cleuziou S, Costantini L (1980) Premiers éléments sur l'agriculture protohistorique de l'Arabie orientale. Paléorient 6(1):245–251

Cleuziou S, Tosi M (2007) In the Shadow of the Ancestors. The Prehistoric Foundations of the Early Arabian Civilization in Oman. Ministry of Heritage and Culture, Sultanate of Oman, Muscat

Córdoba JM (2013) New perspectives about Iron Age and the oasis culture in the Oman peninsula. Two conclusive seasons at al Madam (Sharjah, UAE). Isimu 16:139–151

Córdoba JM, Del Cerro MDC (2005) Archéologie de l'eau dans al Madam (Sharjah, Emirats Arabes Unis). Puits, aflaj et sécheresse pendant l''âge du Fer. Iranica Antiqua 40:515–532

Costa PM (1983) Notes on traditional hydraulics and agriculture in Oman. World Archaeol 14 (3):273–295

Costa PM, Wilkinson TJ (1987) The Hinterland of Sohar, Archaeological Surveys and Excavations within the Region of an Omani Seafaring City. J Oman Stud 9, Ministry of Heritage and Culture, Sultanate of Oman, Muscat

Costantini L (1979) Palaeoethnobotany: identification of the two rows barley and early domesticated sorghum. Archaeology in the United Arab Emirates II-III: 70

Costantini L, Audisio P (2001) Plant and insect remains from the Bronze Age site of Ra's al-Jinz (RJ-2), Sultanate of Oman. Paléorient 26(1): 143–156

Costantini L, Costantini-Biasini L (1986) Palaeobotanical investigations in the Middle East and the Arabian Peninsula. East West 36(4):354–365

Dalongeville R (1999) Mleiha: étude physique et paléoenvironnement. In: Mouton M (ed.) Mleiha I. Environnement, stratégies de subsistance et artisanats. Travaux de la Maison de l'Orient 29, de Boccard, Paris, pp 33–54

Deadman WM (2012) Defining the Early Bronze Age landscape: a remote sensing-based analysis of Hafit tomb distribution in Wadi Andam, Sultanate of Oman. Arab Archaeol Epigr 23(1):26–34

Degli Esposti M (2011) The Excavation of an Early Bronze Age Tower near Salut (Bisyah, Sultanate of Oman): the Iron Age levels. Egitto e Vicino Oriente 34: 189–224

Degli Esposti M, Benoist A (2015) More on Masafi ancestors: the Late Bronze Age site of Masafi-5. Proc Semin Arab Stud 45:57–74

Desruelles S, Fouache E, Eddargach W, Cammas C, Wattez J, Beuzen-Waller T, Martin C, Tengberg M, Cable C, Thornton C, Murray A (2016) Evidence for early irrigation at Bat (Wadi Sharsah, northwestern Oman) before the advent of farming villages. Quatern Sci Rev 150: 42–54. doi: org/10.1016/j.quascirev.2016.08.007

Dutton RW (1989) Aflaj renewal in Araqi: a village case study from Oman. In: Beaumont P, Bonine M, McLachlan K (eds) Qanat, Kariz and Khattara: Traditional Water Systems in the Middle East and North Africa. School of Oriental and African Studies, University of London and Middle East and North African Studies Press, London and Cambridgeshire, pp 237–256

Fleitmann D, Burns SJ, Mangini A, Mudelsee M, Kramers J, Villa I (2007) Holocene ITCZ and Indian monsoon dynamics recorded in stalagmites from Oman and Yemen (Socotra). Quatern Sci Rev 26(1–2):170–188. doi:10.1016/j.quascirev.2006.04.012

Fleitmann D, Matter A (2009) The speleothem record of climate variability in Southern Arabia. CR Geosci 341(8–9):633–642. doi:10.1016/j.crte.2009.01.006

Fouache E, Desruelles S, Eddergach W, Cammas C, Wattez J, Martin C (2012) When was irrigation first used in Bat (Wadi Sharsah, northwestern Oman)? Poster. EGU General Assembly 2012, Held 22–27 April, in Vienna, Austria

Frifelt K (1985) Further Evidence of the Third Millennium BC Town at Bāt in Oman. J Oman Stud 7:89–104

Frifelt K (2002) Bat, a Centre in the Third Millennium Oman. In: Cleuziou S, Tosi M, Zarins J (ed) Essays on the Late Prehistory of the Arabian Peninsula. Serie Orientale Roma XCIII, Istituto Italiano per l'Africa e l'Oriente, Rome, pp 101–110

Giraud J (2007) Reconstitution d'un espace géographique ancien: La province du Ja'alan à l'Âge du Bronze (Sultanat d'Oman). PhD dissertation, University of Paris 1 Panthéon Sorbonne

Giraud J (2009) The evolution of settlement patterns in the eastern Oman from the Neolithic to the Early Bronze Age (6000–2000 BC). CR Geosci 341(8–9):739–749. doi:10.1016/j.crte.2009.03.005

Gros-Balthazard M, Newton C, Ivorra S, Tengberg M, Pintaud J-C, Terral J-F (2013) Origine et domestication du palmier dattier (*Phoenix dactylifera* L.). Revue d'ethnoécologie 4. http://ethnoecologie.revues.org/1524. Accessed 8 Feb 2015

Hastings A, Humphries JH, Meadow RH (1975) Oman in the Third Millennium BC. J Oman Stud 1:1–55

Hogarth DG (1904) The Penetration of Arabia. A record of the development of Western knowledge concerning the Arabian Peninsula. Frederick A. Strokes Company, New York

Kennet D, Velde C (1995) Third and early second-millennium occupation at Nud Ziba, Khatt (U.A.E.). Arab Archaeol Epigr 6:81–99

Kunter M (1981) Bronze und eisenzeitliche Skelettfunde aus Oman. Homo 32:197–210

Laureano P (1998) The Oasis Model: Symbiosis, Intensification, Autopoiesis. In: Pearce M, Tosi M (eds) Papers from the EAA Third Annual Meeting at Ravenna 1997, vol I: Pre- and Protohistory. BAR International Series 717, Archaeopress, Oxford, pp 71–72

Magee P (2014) The archaeology of prehistoric Arabia. Adaptation and Social Formation from the Neolithic to the Iron Age, Cambridge University Press, New York

Méry S (2013) The first oases in Eastern Arabia: society and craft technology in the 3rd millennium BC at Hili, United Arab Emirates. Revue d'ethnoécologie 4. http://ethnoecologie.revues.org/1631. doi:10.4000/ethnoecologie.1631

Mouton M (1999) Mleiha: environnement, stratégies de subsistance et artisanats. In: Mouton M (ed) Mleiha I. Environnement, stratégies de subsistance et artisanats. Travaux de la Maison de l'Orient 29, de Boccard, Paris, pp 265–285

Mouton M, Tengberg M, Bernard V, Le Maguer S, Reddy A, Soulié D, Le Grand M (2012) Building H at Mleiha: new evidence of the late pre-Islamic period D phase (PIR. D) in the Oman peninsula (second to mid-third century AD). Proc Semin Arab Stud 42:205–221

Nash H (2011) Water management: the use of stars in Oman. Society for Arabian Studies Monographs 11, Archaeopress, Oxford

Nelson GC, Lukacs JR, Yule P (1999) Dates, caries, and early tooth loss during the iron age of Oman. Am J Phys Anthropol 108:333–343

Orchard J, Orchard J (2007) The Third Millennium BC Oasis Settlement of Oman and the First Evidence of their Irrigation by Aflaj from Bahla. In: Archaeology of the Arabian Peninsula through the Ages. Proceedings of the International Symposium (7th–9th May 2006). Ministry of Heritage and Culture, Sultanate of Oman, Muscat, pp 143–173

Parker AG, Eckersley L, Smith MM, Goudie AS, Stokes S, Ward S (2004) Holocene vegetation dynamics in the northeastern Rub' al-Khali desert, Arabian Peninsula: a phytolith, pollen and carbon isotope study. J Quat Sci 19(7):665–676. doi:10.1002/jqs.880

Parker AG, Goudie AS, Stokes S, White K, Hodson MJ, Manning M, Kennet D (2006) A record of Holocene climate change from lake geochemical analyses in southeastern Arabia. Quatern Res 66(3):465–476. doi:10.1016/j.yqres.2006.07.001

Pena-Chocarro L, Barrón Lopez E (1999) Plant remains from the site of Mleiha. In: Mouton M (ed.) Mleiha I. Environnement, stratégies de subsistance et artisanats. Travaux de la Maison de l'Orient 29, de Boccard, Paris, pp 63–69

Peyron G (2000) Cultiver le palmier-dattier. Cirad, Montpellier

Potts D (1990) The Arabian Gulf in Antiquity. Volume I : From the Prehistory to the Fall of the Achaemenid Empire. Oxford University Press, NewYork

Potts D (1999) 'The plant for the heart grows in Magan…': redefining Southeastern Arabia's role in ancient Western Asia. Aust Archaeol 48:35–41

Power T, Sheehan P (2011) Bayt Bin 'Ātī in the Qattārah oasis: a prehistoric industrial site and the formation of the oasis landscape of al-'Ain, UAE. Proc Semin Arab Stud 41: 267–282

Power T, Sheehan P (2012) The origin and development of the oasis landscape of al-'Ain (UAE). Proc Semin Arab Stud 42:291–308

Rowley-Conwy P, Deakin W, Shaw CH (1999) Ancient DNA from Sorghum. The evidence from Qasr Ibrim, Egyptian Nubia. In: van der Veen M (ed) The exploitation of plant resources in ancient Africa. Springer, Boston, pp 55–61. doi:10.1007/978-1-4757-6730-8_5

Sanlaville P (2000) Le Moyen-Orient arabe. Le milieu et l'homme. Armand Colin, Paris

Schreiber J (2007) Transformationsprozesse in Oasensiedlungen Omans. Die Vorislamische Zeit am Beispiel von Izki, Nizwa und dem Jebel Akhdar. PhD dissertation, Ludwig-Maximilans University, München

Tengberg M (1998) Paléoenvironnements et économie végétale en milieu aride - recherches archéobotaniques dans la région du Golfe arabo-persique et dans le Makran pakistanais (4ème millénaire av. notre ère - 1er millénaire de notre ère). PhD dissertation, University of Montpellier 2

Tengberg M (2003) Archaeobotany in the Oman peninsula and the role of Eastern Arabia in the spread of African crops. In: Neumann K, Butler A, Kahlheber S (ed) Food, Fuel and Fields: Progress in African Archaeobotany. Africa Praehistorica 15, Heinrich Barth Institut, Köln, pp 229–237

Tengberg M (2012) Beginnings and early history of date palm garden cultivation in the Middle East. J Arid Environ 86:139–147. doi:10.1016/j.jaridenv.2011.11.022

Weisgerber G (1981) Mehr als Kupfer in Oman. Ergebnisse der Expedition 1981. Der Anschnitt 33: 174–263

Weisgerber G (2005) The impact of the dynamics of qanats and aflaj on oases in Oman: Comparisons with Iran and Bahrain. Proceedings of the Internatinales Frontinus-Symposium "Wasserversorgung aus Qanaten – Qanate als Vorbilder im Tunnelbau", 2.-5. Oktober 2003, Walferdange, Luxemburg. Schriftenreihe der Frontinus-Gesellschaft 26: 61–97

Weisgerber G, Yule P (2003) Al-Aqir near Bahla¯' – an Early Bronze Age dam site with planoconvex "copper" ingots. Arab Archaeol Epigr 14: 24–53

Wilkinson JC (1977) Water and Tribal Settlement in South-East Arabia, a study of the Aflāj of Oman. Clarendon Press, Oxford

Willcox G (1995) Some plant impressions from Umm an-Nar island. In: Frifelt K (ed) The Island of Umm An-nar: the Third Millennium Settlement. Jutland Archaeological Society Publications, Aarhus University Press, Aarhus, pp 257–259

Willcox G, Tengberg M (1995) Preliminary report on the archaeobotanical investigations at Tell Abraq with special attention to chaff impressions in mud brick. Arab Archaeol Epigr 6:129

The Oasis of the Chicama Valley: Water Management from the Chimú to the Spaniards (Eleventh to Seventeenth Century AD) on the North Coast of Peru

Camille Clément

Abstract The Chicama Valley is an extremely arid region, irrigated by its eponymous river that forms an oasis. From the eleventh century AD, the Chimú culture developed the irrigation system, extending the limits of this oasis. After the Inca conquest ca. 1470 AD, and especially in 1532 AD with the arrival of the Spaniards, many upheavals (demographic collapse and abandonment of canals) transformed the landscape of the valley. From then on, the exploitation of the Chicama Valley changed dramatically. This raises the question of the management of this oasis by the Chimú, by the Incas and then by the Spaniards. What were the economic policies of land use and allocation of water resources during these three periods? The combination of archaeological data and ethnohistorical information from colonial archives provides an understanding of the processes of economic transformation of this oasis, between the eleventh and seventeenth centuries AD.

Keywords Oasis · Peru · North coast · Chicama · Chimú · Inca · Colonial Period · Irrigation · Agriculture

1 Introduction

The Chicama Valley, located on the north coast of Peru, is an extremely arid region that receives only a few millimetres of rainfall per year. Nevertheless, this aridity is compensated by the presence of the Río Chicama, the eponymous river of the valley that descends from the peaks of the Andes. On the banks of this river, vegetation grows, forming a welcoming oasis where many human cultures, such as the Chimú, have succeeded over the millennia. From the eleventh century AD, the Chimú

C. Clément (✉)
University Paris 1 Panthéon-Sorbonne, UMR 8096 ARCHAM, Paris, France
e-mail: camclement73@hotmail.fr

C. Clément
Maison René-Ginouvès, 21 Allée de L'Université, 92023 Nanterre Cedex, France

E. Lavie and A. Marshall (eds.), *Oases and Globalisation*,
Springer Geography, DOI 10.1007/978-3-319-50749-1_5

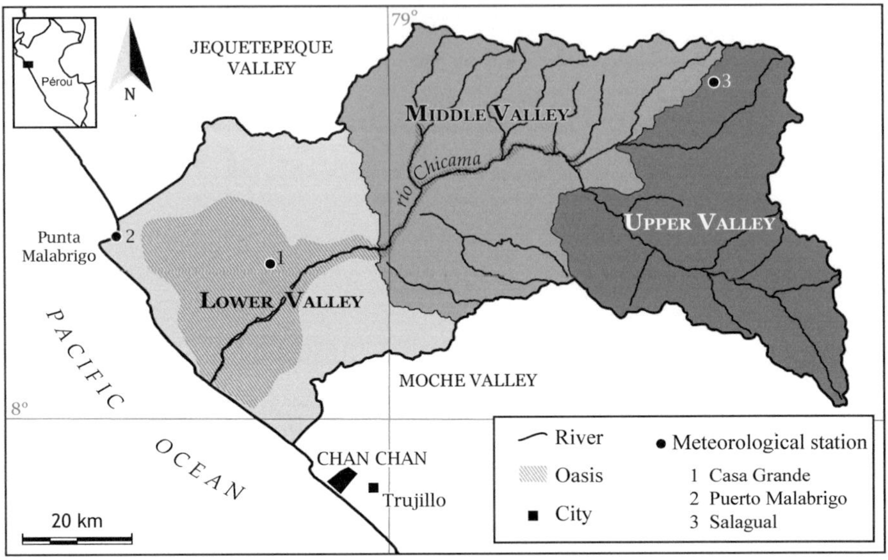

Fig. 1 Map of the Chicama Valley (data from IGN Peru and ONERN 1973)

initiated the expansion of the oasis of Chicama, through irrigation, farming a large area (Fig. 1). The Chimú Kingdom, whose political and social organisation depended on canals, was conquered by the Incas ca. 1470 AD, and then in 1532 AD by the Spaniards. The political, economic and social upheavals that resulted from these two successive conquests, such as a demographic collapse and the abandonment of some irrigation canals, had a direct impact on the oasis of Chicama and its exploitation.

Thus, we will analyse the changes and transformations of the economic management modes of this oasis, between the eleventh century, with the beginning of the Chimú culture in the valley, and the early seventeenth century, during the first hundred years after the Spanish conquest, when the old traditions from pre-hispanic times still survived. After presenting the geographical setting, the Chicama Valley and its oasis, and the history of the region through the Chimú, Inca and Colonial Periods, we discuss the methodology and the results of our ethnohistorical and archaeological studies. Finally, we examine all of these data to understand the evolution of the management and exploitation of this oasis during different periods.

2 The Chicama Valley and Its Oasis

Bordered by the Andes and the Pacific Ocean, the north coast of Peru, where the oasis of Chicama is located, receives only a fraction of the rainfall coming from the east. Other factors, including the temperature inversion caused by the cold waters of the Humboldt Current in the Pacific Ocean, also contribute to the aridity of this region (Orbegoso 1987). The few millimetres of annual rainfall recorded in the Chicama Valley at low altitude (5 mm at Puerto Malabrigo and 20 mm at Casa Grande; Fig. 1; ONERN 1973) testify to this desert climate. However, the many rivers that flow down the western slopes of the Andes break the aridity and create several oases, such as that of Chicama. The role of these rivers is central as they are often the only source of fresh water on the coast.

The Chicama Valley, bounded in the north by the Jequetepeque Valley (and the Quebrada Cupisnique) and in the south by the Moche Valley, can be divided into three regions: lower, middle and upper (Fig. 1). The lower valley is a broad coastal plain, strewn with small rocky elevations and bordered by the Pacific Ocean and the foothills of the Andes. Today, most of the oasis of Chicama is concentrated in this lower part (ONERN 1973). The middle area is characterised by a hilly landscape, although the bottom of the valley is still relatively wide. The low-altitude (below 1200 m.a.s.l.) desert climate prevents the development of vegetation on the mountain slopes, which are only covered with some cacti (ONERN 1973). However, the various streams, such as the Río Chicama and its tributaries, favour the growth of plants and so the oasis is extended near the rivers. Lastly, in the upper valley, on the hill top rising over 4000 m, the climate is wetter. In fact, this area receives more than 1000 mm of annual rainfall (1130 mm in Salagual; Fig. 1; ONERN 1973), which gives rise to many rivers, including the Río Chicama.

The use of the term "oasis" for the valleys of the Peruvian coast is quite recent. It is found in the reports of scientific explorers of the nineteenth century, like Wiener (2010 [1880]: 105), but the Spanish chroniclers of the sixteenth and seventeenth centuries, such as Garcilaso de la Vega (1982 [1617]) and Cieza de León (2005 [1553]), never used this term. In the case of the Chicama Valley, is it relevant to speak of an "oasis"? There is no doubt that the desert coast contrasts with the plants of the riverbanks, but the vegetation, such as *algarrobals* and *sapotals*—mostly composed of shrubs and trees—survives in areas relatively far from rivers. The landscapes of the arid coast do not always lack flora. Moreover, the oasis of the Chicama Valley is highly anthropised since irrigation helped to cultivate a large part of the plain, extending the area of the oasis.

The Chicama oasis can therefore be defined as a river oasis, enlarged through an extensive network of canals and fields, contrasting this green landscape with the surrounding desert. It is interesting to note that, on one hand, the oasis of Chicama is separated from those of Moche and Jequetepeque by very arid regions but, on the other hand, it is "connected" by its river to the wetter areas of the hills. In this sense, the oasis of Chicama is not completely isolated, as could be the case for a Saharan oasis.

The transformation of the natural oasis by irrigation and its exploitation by human beings show the close relationship maintained with this particular environment by populations who have settled in the Chicama Valley for millennia.

3 From the Chimú Culture to Colonial Times

From the eleventh century AD, the Chimú culture developed in the south of the Chicama Valley, in the Moche Valley, on the site of Chan Chan (Fig. 1). Many archaeologists have shown interest in this major Chimú centre (Rodríguez 1968; Moseley and Day 1982; Topic and Moseley 1983; Moseley and Cordy-Collins 1990; Pillsbury and Leonard 2004; Campana 2006, etc.) and have described the different buildings that composed the twenty square kilometres of Chan Chan. They identified ten monumental adobe buildings, called *ciudadelas*, which arc divided into large ceremonial plazas, corridors and courtyards, containing U-shaped structures and storerooms (Day 1982). The funeral platforms of these compounds could have contained the burials of the deceased Chimú rulers (Conrad 1982). These *ciudadelas* probably played the role of ceremonial and administrative palaces, transformed into funeral mausoleums (Moseley and Day 1982; Campana 2006). In addition to these great compounds, intermediary buildings, residences of the elite members (Klymyshyn 1982), small irregular agglutinated rooms (SIAR), and housing specialist craftsmen (Topic 2009) formed the urban network of this site.

The Chimú gradually spread over hundreds of kilometres of coastline, unifying a vast territory, managed through administrative centres, such as Quebrada del Oso in the Chicama Valley (Keatinge 1974) and Farfan in the Jequetepeque Valley Mackey (1987; Fig. 2). The architecture of these buildings is directly inspired from the *ciudadelas* of Chan Chan: courts, corridors, U-shaped structures, and sometimes storerooms are found inside the rectangular compounds of these administrative centres. Therefore, Mackey (1987) proposes an administrative hierarchy with, at the top, the capital palaces, then the major administrative centres, and finally the rural administrative centres, smaller in size. As these rural administrative centres were generally located near canals and agricultural fields, Keatinge (1974) assumed that they were used for irrigation management, an essential task for the Chimú economy based primarily on agriculture. Pozorsky (1979) documented the consumption and cultivation of corn, squash, beans, sweet potatoes, peppers, gourds and cotton, to name only a few species. The presence of some Chimú administrative centres in the oasis of Chicama could help understand how this area was exploited, but all the residential sites, where people working in the oasis lived, must also be taken into account.

Around 1470 AD, the Inca army subdued the Chimú Kingdom and incorporated it into the Inca Empire, the Tawantinsuyu (Rowe 1948). The impact of this conquest remains difficult to perceive through archaeological records and in the Chicama Valley. Some sites, like Chiquitoy Viejo, a *tambo* (roadhouse) studied by Conrad (1977) and PV22-23, in the north of the Chicama Valley (Chauchat et al. 1998; Fig. 3), show the Inca interest in communications. However, identifying the Inca policy in the Chicama oasis is difficult with the current state of knowledge.

Sixty years later, in 1532 AD, the Spaniards conquered the Inca Empire and installed a colonial regime. To this end, they founded cities like Trujillo, in the

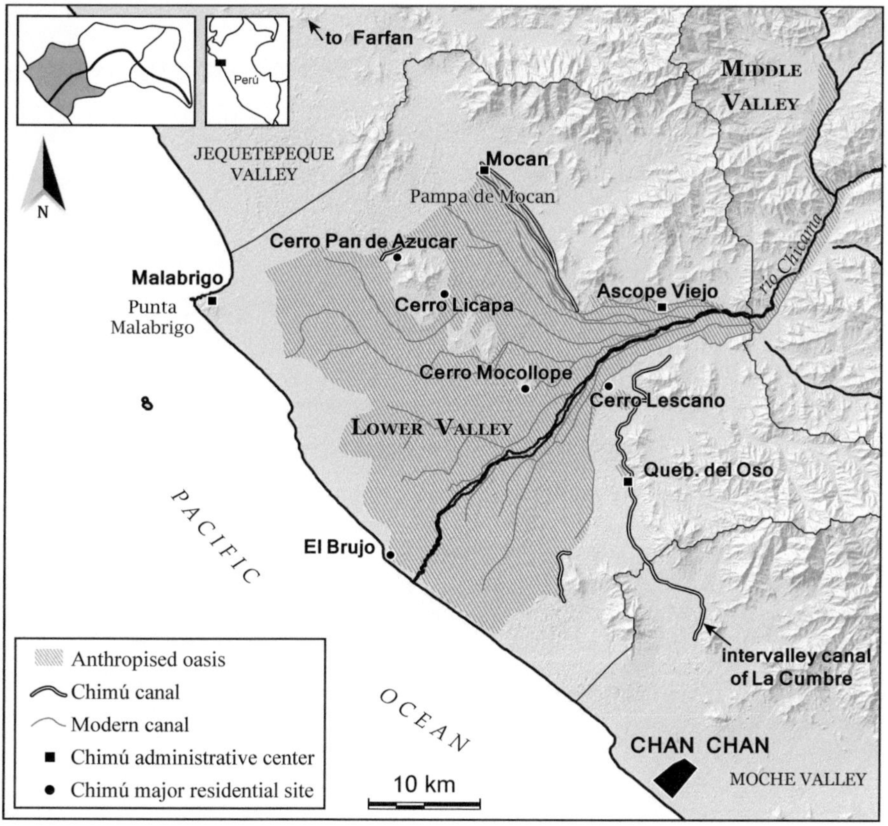

Fig. 2 Map of the oasis of Chicama during the Chimú Period (eleventh century to fifteenth century; data from IGN Peru and Aster GDEM)

Moche Valley near Chan Chan, created *encomiendas*, *repartimientos* and haciendas, to grow wheat or sugar cane, and gathered the indigenous declining population into *reducciones* (Duviols 1971: 248). During the sixteenth century AD, the oasis of Chicama and its people underwent profound changes.

The management of this largely anthropised oasis by the Chimú, and then by the Inca and the Spaniards, has evolved over time. What economic policies of exploitation of land and allocation of water resources were put in place? What original pre-hispanic traditions persisted into the Colonial Period, despite the upheavals of the Spanish conquest? We seek to understand the evolution of the oasis of Chicama by taking into account not only the changes but also the persistence of exploitation and social organisation practices in this particular space.

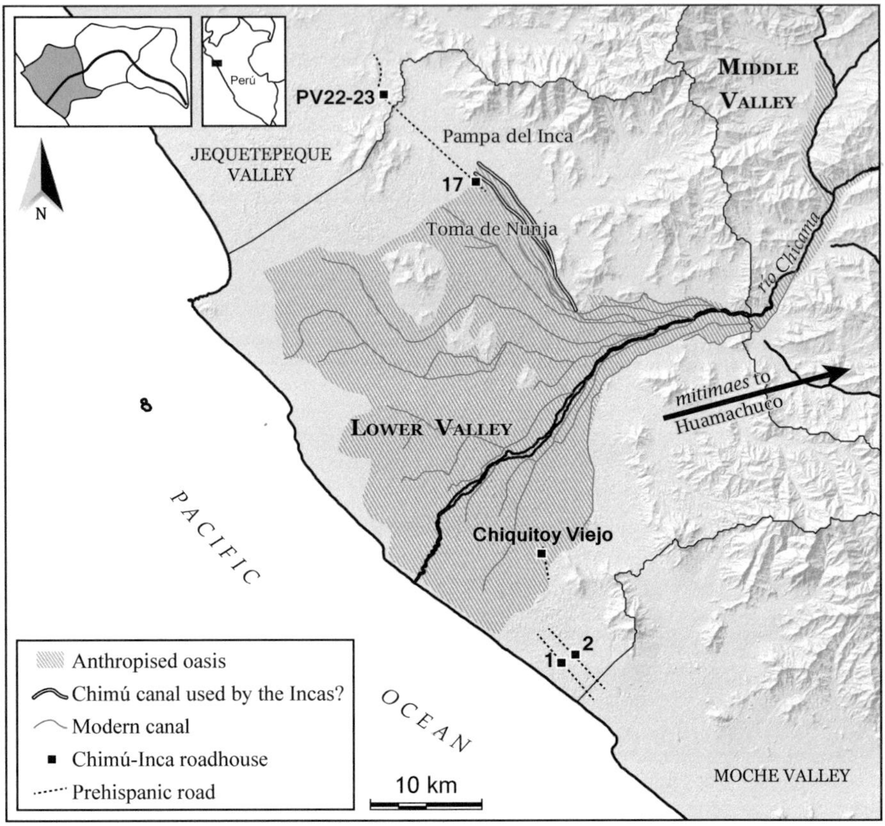

Fig. 3 Map of the oasis of Chicama during the Inca Period (fifteenth century to sixteenth century; data from IGN Peru and Aster GDEM)

4 Studies in the Chicama Valley

The study of these transformations involves data about both pre-hispanic times and the Colonial Period. To understand this process, it is essential to use an innovative multidisciplinary approach including ethnohistory and archaeology.

The situation of the Chicama Valley in the sixteenth century is reflected in the works of ethnohistorians such as Netherly (1977, 1984), Ramirez (1995) and Zevallos Quiñones (1990, 1993). We also consulted documents from the Archivo General de las Indias in Spain, and the Archivo Regional de La Libertad in Peru, for a better understanding of the daily lives of the native population in this oasis, after the Spanish conquest. From all the collected data, it is possible to determine approximately the alleged territories of the local communities of Chicama. However, to observe the oasis during the Chimú and Chimú–Inca Periods, the archaeological data are essential.

Surveys by Chauchat et al. (1998), Leonard and Russell (1992) and Krzanowski (1985) document the occupation of the region during the Late Intermediate Period (900–1470 AD). Studies about Chimú irrigation by Kus (1972), Watson (1979) and Farrington (1983) demonstrate their mastery of hydrology and provide some information on the irrigation network of the oasis. In addition, Google Earth satellite images help to identify more than 400 potential archaeological sites in the lower valley, including many canals, and to draw up draft plans of the vestiges.

Nevertheless, these ethnohistorical, bibliographic and satellite data must be supplemented by field surveys. Between 2010 and early 2013, we carried out several expeditions in 23 zones of the valley, from the desert coast to the hills, along the banks of rivers, identifying a total of 167 archaeological sites (Clément 2015). In addition to analysing the ceramic material in order to date the main phases of occupation of the sites, we also studied the architecture of the various conserved buildings. Some of these sites are mere concentrations of shards scattered on the surface, weak evidence of human activity in the past, while others, in contrast, consist of the ruins of buildings that could correspond to villages. Finally, we were also interested in the ancient roads, some long isolated walls and, of course, the pre-hispanic canals and fields, witnesses of the farming in the oasis of Chicama, centuries ago.

Of these 167 sites, 49 were occupied during the Chimú Period and only 3 during the Chimú–Inca Period. Chimú sites are mostly located near rivers or major irrigation canals and are therefore often related to water (Clément 2014). Some of them, like those of Mocan and Malabrigo (Pozorski 1987; Chauchat et al. 1998), are Chimú administrative centres, whose architecture is directly inspired from the *ciudadelas* of Chan Chan.

Others, such as the site of Cerro Lescano (Fig. 2), are composed of many residential structures built with stones. In these large villages, there are generally imposing buildings with a complex architecture, divided into many parts. Only members of the elite could order the construction of such buildings, and it is reasonable to think that they were residences for high-ranking people (Pillsbury and Leonard 2004). These elite did not appear to be directly linked to the authority of Chan Chan, as these large and complex structures were not derived from the *ciudadelas* of the capital, unlike the administrative centres. As demonstrated by Netherly (1977, 1984) for the Colonial Period, there was probably a small local Chimú nobility who directed the communities of the oasis.

The Chimú canals associated with agricultural fields are mostly located in the desert marginal areas of the coastal plain. Their larger size requires a considerable investment of time and labour and shows the importance of the hydraulic system during the Chimú Period. Finally, we can reconstruct the irrigation system in the currently cultivated areas, since most of the modern canals reuse the routes of the old pre-hispanic ones, as Netherly (1977, 1984) supposed. The irrigation network that once watered the oasis of Chicama emerges through these many clues.

The results of our survey, supplemented by the works of other archaeologists and the survey of satellite images, and associated with information from ethnohistorical texts from the sixteenth and early seventeenth centuries, enable us to understand the evolution of the exploitation of the oasis of Chicama, from the Chimú to the Colonial Period.

5 Interpretations: Exploitation of the Oasis of Chicama

As mentioned above, without anthropogenic infrastructures, the oasis is naturally confined to the banks of the Río Chicama. However, due to irrigation, farming could be developed further, and so the anthropised oasis today extends over a wide area. The canals, and the sites associated with them, thus appear as key indicators for understanding the evolution of the oasis and its management.

5.1 Exploitation by the Chimú

The Chimú extended the hydraulic system, pushing the limits of the oasis at its peak, especially in the north, into the Pampa de Mocan (Watson 1979) and to the east, with the inter-valley canal of La Cumbre, which diverted water from the Río Chicama towards the Moche Valley (Kus 1972). Today, the fields of these marginal areas are abandoned and deserted as the cultivated area is smaller than during the Chimú Period. The agricultural production may have been used to feed a growing population in the Chicama Valley or the numerous inhabitants of the nearby capital, Chan Chan.

To manage the resources of the oasis, both water and land, the Chimú built rural administrative centres, such as Quebrada del Oso (Keatinge 1974), Mocan (Pozorski 1987: 116) and Ascope Viejo (Leonard and Russell 1992), all located close to major canals and cultivated areas (Fig. 2). This association supports the hypothesis initially formulated by Keatinge (1974) about their function of controlling the irrigation.

The distribution of water from the canals to the different plots of the fields in turn, described by Garcilaso de la Vega (1982 [1617]), involves training and a major control to avoid abuse. In fact, in the Inca Empire, an official, named *cillquia*, had to ensure the proper distribution of water among the various users (Guaman Poma de Ayala 1980 [1583–1615]). Although ethnohistorical sources tell us nothing about the existence of this system for the Chimú, it is possible that rural administrative centres housed people responsible for this fair distribution of water from the canals to the fields. At the site of Mocan (Fig. 2), a highly complex network of canals, with many ramifications and sometimes meetings of canals, can be observed. The water could therefore take several paths to reach a field and had to be guided with precision by competent people. Thus, agricultural expansion in the oasis of Chicama was made possible by a particular management of water resources in some Chimú administrative centres. However, this administration began upstream in the middle valley where the Chimú were installed to control areas of confluences and to observe the growth of the flow of the Río Chicama (Clément 2014).

A fourth Chimú administrative centre, named Malabrigo (Fig. 2), is somewhat distant from the oasis. Established in the Punta Malabrigo, overlooking a large sheltered bay, this site seems to have turned towards fishing rather than agriculture. Yet, the guano, a natural fertiliser produced by seabird droppings, accumulated in the Punta Malabrigo, near this site, could also have once been used to improve the cultivated land (Kubler 1948). The contribution of this fertiliser probably had a positive impact on agricultural production in the oasis of Chicama.

These four administrative centres are, at first sight, closely linked to the oasis and farming. However, their location on the margins of the oasis leads us to rethink, in part, their role and importance.

The site of Mocan is located at the end of the irrigation system that extends to the northwest, while Quebrada del Oso is established near the last fields watered by the inter-valley canal in the Chicama Valley (Fig. 2). From these centres, some Chimú functionaries could actually check for water at the end of the irrigation system, but they could not control the distribution of water or supervise the intake of the canals. Only the site of Ascope Viejo (Leonard and Russell 1992) was ideally located to manage irrigation, since it was close to the intakes of some major Chimú canals (Fig. 2). Because of their peripheral and marginal positions, these four administrative centres would not enable the effective control of the kilometres of canals and hectares of fields that constitute the oasis of Chicama.

However, the great residential, administrative and ceremonial sites, such as Cerro Pan de Azucar (Mercado 2002), Cerro Licapa (Netherly 1977), Cerro Mocollope (Leonard and Russell 1992), El Brujo (Tate 2007) and Cerro Lescano (Fig. 2), probably inhabited by members of the local elite, are located in the heart of the coastal plain and the oasis. From these sites, it was possible to manage the irrigation and the cultivated land extending around them. This administrative task had to be led by high-ranking local rulers who lived in these sites, while the population, mainly composed of farmers according to Netherly (1977), worked in the fields.

Therefore, there were two types of management operating in parallel; the marginal areas of the oasis, newly cultivated by the Chimú, were under the indirect control of the leaders of Chan Chan, via administrative centres, while the heart of the oasis was in the hands of local elite members of Chicama, who managed and operated these vast fertile lands from some major villages.

5.2 Exploitation by the Inca

After the period of extension of this oasis by the Chimú, the situation changed with the Inca conquest. Unfortunately, the remains of the Chimú–Inca Period, mostly unknown, do not provide a complete understanding of the exploitation of the coastal plain of Chicama. The three Chimú–Inca sites we identified are installed along roads and may have been roadhouses, similar to the site of Chiquitoy Viejo (Conrad 1977) and PV22-23 (Chauchat et al. 1998; Fig. 3). According to Rowe (1948), the Chimú rulers of Chan Chan lost their power after the Inca conquest, and we can assume that the marginal areas of Chicama they controlled through administrative centres were abandoned, as no Chimú–Inca shards have been found in Quebrada del Oso and the fields of Mocan.

Yet, the presence of an Inca roadhouse in the Pampa de Mocan (site n°17; Fig. 3), south of the Chimú fields, raises questions about the continuity of farming under the Inca domination. The toponyms of the surrounding area, such as the "Pampa del Inca," which extends to the north, or "toma de Nunja," associated, according to Zevallos Quiñones (1993: 54), with the "land of the sun," suggest that the Incas settled in this region. Indeed, the "lands of the sun" were managed by the Inca clergy to worship the sun god Inti (Moore 1973: 24), and their mention might indicate the trace of the Incas. Could the Incas have maintained the Mocan zone under cultivation? Garcilaso de la Vega (1982 [1615] t.2: 65) provided data about the Inca system of land distribution. This author indicates that the Incas were seeking to gain new lands to increase the production, without despoiling the newly conquered communities who often conserved their fields more or less intact. With the recovery of the marginal cultivated areas of Mocan, formerly under the control of the rulers of Chan Chan, the Incas would have enjoyed the many hydraulic infrastructures built during the Chimú period, leaving the local communities to exploit the heart of the oasis. Yet, archaeological data are insufficient to demonstrate the continuity of farming in Mocan, during the Inca Period.

Besides, the texts of the Colonial Period mention the sending of *mitimaes* (settlers) from the Chicama Valley towards the Huamachuco area further east (Fig. 3). Pedro Xalcaguaman, ruler of the *mitimaes* in the early seventeenth century, says that the Incas ordered his grandfather, Augustin Chumbinamo, to go with his community to the mountains, perhaps during the reign of Huayna Capac (ARL leg. 266 exp. 3068 fol. 32r; Espinoza Soriano 1974). These population movements weakened the demographics in the oasis of Chicama. Workers, inevitably fewer, could not exploit all the land that was irrigated in Chimú time.

Data are lacking to explain in detail how the oasis of Chicama was occupied and managed during the Chimú–Inca Period. Although some marginal agricultural areas appear to have been abandoned, others, like Mocan, could have been maintained under cultivation by the Incas, who replaced the authority of the Chimú rulers of Chan Chan. The population movements also contributed to reducing the area of the oasis, given the lack of manpower.

5.3 The Colonial Exploitation: Upheaval and Survival

With the arrival of the Spaniards on the north coast of Peru, the landscape of the oasis of Chicama completely changed as a result of the restructuring of the indigenous society, the demographic collapse suffered by the native population and the introduction of new crops, exogenous domesticated animals, and technologies, like mills (Piel 1975; Cook 1981: 143; Netherly 1977; Castañeda 1996).

The Spaniards tried to reshape the indigenous society to control the population better. In the sixteenth century, the natives of Chicama were initially placed under the responsibility of two *encomenderos*, Don Diego de Mora and Don Francisco Fuentes, who took part of their produce as tribute (Zevallos Quiñones 1990). In 1566, Dr. Cuenca grouped various hamlets into a few villages, the *reducciones*, such as the villages of Paijan, Chocop, Magdalena de Cao and Santiago de Cao (AGI Justicia 457; Fig. 4). These measures facilitated not only the conversion of the inhabitants of the valley to Catholicism, but also their economic control (Duviols 1971).

In the sixteenth century, the population of the north coast and of the oasis of Chicama suffered a dramatic demographic collapse, aggravated by the social and economic changes, and especially by the epidemics that decimated the indigenous

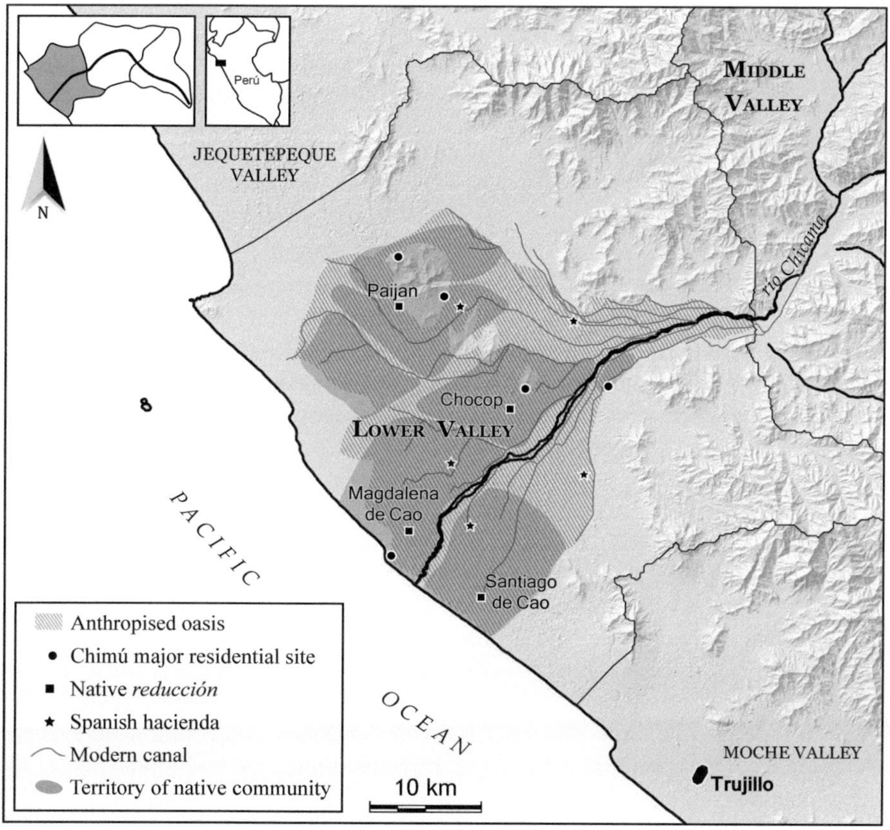

Fig. 4 Map of the oasis of Chicama during the Colonial Period (sixteenth century to seventeenth century; data from IGN Peru and Aster GDEM)

communities, as in 1578 (ARL leg. 266 exp. 3064). The population estimates are difficult to verify, but all authors (Cook 1981; Netherly 1977; Ramírez 1995) agree on the high rate of mortality and the consequences of this decline on the irrigation system. Due to a labour shortage, canal maintenance and exploitation of the extensive fields of the oasis of Chicama could not continue and many areas had to be abandoned, particularly in the marginal zones irrigated by the Chimú, a few centuries previously (Fig. 4). Like for the estimation of the demographic collapse, it is very difficult to assess the reduction in the cultivated area of the oasis.

The landscape was also transformed by the creation of Spanish haciendas dedicated to raising cattle and goats and the cultivation of wheat and sugar cane (Feyjoo de Sosa 1984 [1763]; Castañeda 1996). Sugar cane gradually covered a large area of the oasis and created social problems about water distribution. The archives (ARL leg. 169 exp. 594; ARL leg. 270 exp. 3335) mention conflicts over the misappropriation of water by Spaniards to grow sugar cane, a plant with great water needs, depriving indigenous villages downstream of the canals. From these haciendas, the Spaniards sent wheat flour to the markets of Trujillo and Lima, and exported sugar outside of Peru. The oasis of Chicama was then integrated into a large trade network (Castañeda 1996).

The social and economic transformations that affected the oasis of Chicama during the early Colonial Period were profound. The demographic collapse and the abandonment of cultivated areas, the Spaniards' *reducciones* policy and the newly established haciendas permanently changed the landscape of this oasis. Even today, almost five centuries after its introduction, sugar cane covers most of the irrigated land in Chicama (ONERN 1973).

Despite the upheavals caused by the Spanish exploitation, certain continuities in the structures of the indigenous society and in the agricultural practices can also be observed through the texts of the archives.

First, Netherly (1977, 1984), Zevallos Quiñones (1990) and Ramirez (1995) studied the role of the indigenous leaders of Chicama. Each community, led by its *curaca*, cultivated its land and maintained its canals for its subsistence and to pay the tribute (Netherly 1984; AGI Justicia 457; AGI Justicia 458). The *curacas* played an important role in agricultural activities: their presence was necessary to supervise the native population's activities and to offer corn beer (*chicha*) to the workers (AGI Justicia 457), in accordance with the Andean principle of reciprocity, inherited from pre-Hispanic times (Franklin 1992).

These people, whose role was essential for farming in the oasis of Chicama, presented themselves as the descendants of members of the local Chimú elite (Netherly 1977), who probably lived in the large villages, such as Cerro Pan de Azucar, Licapa, Cerro Mocollope, El Brujo and Cerro Lescano (Figs. 2 and 4). We note that these great Chimú sites fall inside the territories of the various indigenous communities, determined on the basis of ethnohistorical data and toponyms for the sixteenth century. These sites could therefore correspond to the major centres of the communities of the oasis during the Chimú Period. Their inhabitants were later resettled by the Spaniards in *reducciones*, often located close to these important sites, as has been documented for the *reducción* of Magdalena de Cao and the site of El Brujo (Mujica and Hirose Maio 2007).

Thus, the ethnohistorical data demonstrate continuity, despite the profound changes that marked the sixteenth century. Native communities, led by their *curacas*, continued to cultivate their land, according to traditional Andean principles such as reciprocity and the distribution of *chicha*. The people of the oasis, fewer every year, had to face competition and economic pressure from the Spaniards who settled in the oasis. Lastly, it should be noted that the area of the oasis was reduced, due to the abandonment of canals and marginal lands that have remained deserted until today.

6 Conclusion

In conclusion, we can trace the evolution of the occupation and exploitation of the oasis, which extended to its maximum during the Chimú Period. This oasis is largely based on human infrastructures, such as irrigation canals, and therefore has an anthropogenic nature, which makes this region particularly responsive to changes in human societies.

The expansion of the oasis of Chicama was probably the culmination of a long process that began in previous periods, with the construction of canals, larger each time and irrigating still further (Golte 1980). In the Moche Valley, Moseley et al. (1983) documented the creation of new canals, and also the partial collapse of the irrigation system, at the end of the Chimú Period. These processes probably occurred in other valleys of the north coast. In a region with a desert climate, such agricultural development requires careful management of water resources, which could be carried out from the Chimú administrative centres, such as Quebrada del Oso, Mocan and Ascope Viejo, symbolising the central authority of Chan Chan, the Chimú capital (Keatinge 1974; Pozorski 1987). However, the location of these sites did not allow an effective control over the whole oasis and all the canals. Thereby, the local elite residing in large sites, probably centres of the communities of Chicama, could manage the exploitation of the vast fields, in the heart of the oasis, whereas state functionaries acted from administrative centres in the peripheral zones.

During the Inca Period, the marginal cultivated areas were probably abandoned —although the Incas might have maintained the exploitation in Mocan—but the reduction in the cultivated acreage was most pronounced following the demographic collapse of the sixteenth and seventeenth centuries. As the oasis was reduced, the Spaniards settled there, cultivating new plants, such as wheat and sugar cane, and opening the valley towards a large trade network (Castañeda 1996). Despite the economic and social changes brought about by the Spaniards, such as the *reducciones* and the *encomiendas*, the native population continued to cultivate the oasis, according to ancient pre-hispanic principles, such as reciprocity. The *curacas*, descendants of the local Chimú elite, led their communities while integrating themselves into the new political machinery of the colonial system.

References

Campana Delgado C (2006) Chan Chan del Chimo: estudio de la ciudad de adobe más grande de América antigua. Editorial Orus, Lima

Castañeda Murga J (1996) Notas para una Historia de la ciudad de Trujillo del Perú en el siglo XVII. In: Tomeada H, Millones L (eds) La Tradición Andina en Tiempos Modernos. National Museum of Ethnology, Osaka, pp 159–190

Chauchat C, Gálvez Mora CA, Briceño Rosario J, Uceda Castillo S (1998) Sitios arqueológicos de la zona de Cupisnique y margen derecha del valle de Chicama. Instituto Nacional de Cultura La Libertad, Trujillo and Institut Français d'Etudes Andines, Lima

Cieza de León P (2005) (1553) Crónica del Perú, el Señorío de los Incas. Fundación Biblioteca Ayacucho, Ayacucho

Clément C (2014) L'eau: Une ressource cruciale pour la culture chimú dans la vallée de Chicama (côte Nord Du Pérou). Archéo Doct 6:57–75

Clément C (2015) Paysage socioculturel et architecture dans la culture chimú. BAR International Series 2748, Paris Monographs in American Archaeology 42, Archaeopress, Oxford

Conrad GW (1977) Chiquitoy Viejo: an inca administrative center in the Chicama valley, Peru. J Field Archaeol 4(1):1–18

Conrad GW (1982) The burial platforms of Chan Chan: some social and political implications. In: Moseley ME, Day KC (eds) Chan Chan: Andean Desert City. University of New Mexico Press, Albuquerque, pp 87–117

Cook ND (1981) Demographic collapse, Indian Peru, 1520–1620. Cambridge University Press, Cambridge

Day KC (1982) Storage and labor service: a production and management design for the Andean area. In: Moseley ME, Day KC (eds) Chan Chan: Andean desert city. University of New Mexico Press, Albuquerque, pp 333–349

Duviols P (1971) La lutte contre les religions autochtones dans le Pérou colonial «L'extirpation de l'idolâtrie» entre 1532 et 1660. Institut Français d'Etudes Andines, Lima

Espinoza Soriano W (1974) Los señoríos étnicos del valle de Condebamba y provincia de Cajabamba. Etnohistoria de las Huarancas de Llucho y Mitmas. Siglos XV–XX. Anales Científicos de La Universidad Nacional Del Centro Del Perú 3:5–369

Farrington IS (1983) The design and function of the intervalley canal: comments on a paper by Ortloff, Moseley, Fledman. Am Antiquity 48(2):360–375

Feyjoo de Sosa M (1984) [1763] Relación Descriptiva de la Ciudad, y Provincia de Truxillo del Perú, con noticias exactas de su estado político. Fondo del Libro—Banco Industrial del Perú, Madrid

Franklin Pease GY (1992) Curacas, reciprocidad y riqueza. Pontificia Universidad Católica del Perú, Lima

Garcilaso de la Vega I (1982) [1617] Commentaires royaux sur le Pérou des Incas. Edition La Découverte, Paris

Golte J (1980) Notas sobre la agricultura de riego en la costa peruana. Allpanchis 14:57–67

Guaman Poma de Ayala F (1980) (1583–1615), El primer nueva corónica y buen gobierno. Siglo XXI, Mexico

Keatinge RW (1974) Chimu rural administrative centres in the moche valley, Peru. World Archaeology 6(1):66–82

Klymyshyn AM (1982) Elite compounds in Chan Chan. In Chan Chan: Moseley M, Day KC (eds) Andean desert city. University of New Mexico Press, Albuquerque, pp 119–143

Krzanowski A (1985) Implicaciones demográficas del patrón de asentamiento prehispánico en los Andes. El caso del valle Alto Chicama, Perú. J Soc des Américanistes 71:79–96

Kubler G (1948) Towards absolute time: guano archaeology. Mem Soc Am Archaeol 4:29–50

Kus JS (1972) Selected aspects of irrigated agriculture in the Chimu Heartland, Peru. Ph.D. dissertation, University of California

Leonard BL, Russell GS (1992) Informe preliminar: proyecto de reconocimiento arqueológico del Chicama, resultados de la primera temporada de campo, 1989. Lima

Mackey CJ (1987) Chimu administration in the provinces. In: Haas J, Pozorski SG, Pozorski T (eds) The origins and development of Andean state. Cambridge University Press, Cambridge, pp 121–129

Mercado Rodríguez CR (2002) Estudio de un asentamiento chimú en Cerro Pan de Azucar. Tesis de licenciatura, Universidad Nacional de Trujillo, Valle de Chicama

Moore SF (1973) Power and Property in Inca Peru. Greenwood Press, Westport

Moseley ME, Cordy-Collins A (1990) The northern dynasties: kingship and statecraft in Chimor: a symposium at Dumbarton Oaks, 12th and 13th October 1985. Dumbarton Oaks Research Library and Collection, Washington, D.C

Moseley ME, Day KC (1982) Chan Chan, Andean Desert City. University of New Mexico Press, Albuquerque

Moseley ME, Feldman RA, Ortloff CR, Narváez AV (1983) Principles of Agrarian collapse in the cordillera negra, Peru. Ann Carnegie Museum 52:299–327

Mujica BE, Hirose Maio E (2007) El Brujo. Huaca Cao, centro ceremonial en el Valle de Chicama. Fundación Wiese, Lima

Netherly PJ (1977) Local Level Lords on the North Coast of Peru. Ph.D. dissertation, Cornell University

Netherly PJ (1984) The management of late Andean irrigation systems on the North Coast of Peru. Am Antiq 49(2):227–254

ONERN (1973) Inventario, evaluación y uso racional de los recursos naturales de la costa-Cuenca del Río Chicama. Oficina Nacional de Evaluación de Recursos Naturales, Lima

Orbegozo Rodríguez E (1987) Geografía de La Región de La Libertad. CONCYTEC, Lima

Piel J (1975) Capitalisme agraire au Pérou. Premier volume. originalité de la société agraire péruvienne au XIXe siècle. Institut français d'études andines, Anthropos Editorial, Lima

Pillsbury J, Leonard BL (2004) Identifying Chimú palaces: Elite residential architecture in the late intermediate period. In: Evan ST, Pillsbury J (eds) Palaces of the ancient new world. A symposium at Dumbarton Oaks 10th and 11th October 1998. Dumbarton Oaks Research Library and Collection, Washington, D.C, pp 247–298

Pozorski SG (1979) Prehistoric diet and subsistence of the Moche valley, Peru. World Archaeol 11(2):163–184

Pozorski T (1987) Changing priorities within the Chimu state: the role of irrigation agriculture. In: Haas J, Pozorski SG, Pozorski T (eds) The origins and development of Andean state. Cambridge University Press, Cambridge, pp 111–120

Ramírez SE (1995) De pescadores y agricultores: Una historia local de la gente del valle de Chicama antes de 1565. Bulletin de l'IFEA 24(2):245–279

Rodríguez Suy Suy VA (1968) Chan Chan: ciudad de adobe. observaciones sobre su base ecológica. In Actas y Memorias del XXXVII congreso internacional de Americanistas Argentina, 1966. Depto. de Publicaciones Científicas, Buenos Aires, pp 133–154

Rowe JH (1948) The kingdom of Chimor. Acta Am 6:26–59

Tate JP (2007) The late horizon occupation of the El Brujo site complex, Chicama Valley, Peru. Ph.D. dissertation, University of California

Topic JR (2009) Domestic economy as political economy at Chan Chan, Peru. In: Manzanilla LR, Chapdelaine C (eds) Domestic life in prehispanic capitals. A study of specialization, hierarchy and ethnicity. University of Michigan Museum of Anthropology, Ann Arbor, Michigan, pp 221–242

Topic JR, Moseley ME (1983) Chan Chan: a case study of urban change in Peru. Nawpa Pacha: J Andean Archaeol 21:153–182

Watson RP (1979) Water control and land use on the arid North Coast of Peru: prehispanic agricultural systems in the Chicama valley. M.A. Thesis, University of Texas at Austin
Wiener C (2010) (1880), Voyage au Pérou et en Bolivie (1875-1877. Ginkgo éditeur, Paris
Zevallos Quiñones J (1990) Pretendientes a la encomienda de Chicama en 1583. Boletín del Instituto Riva-Agüero 17:373–381
Zevallos Quiñones J (1993) Fuentes para el estudio de la lengua Quingnam 1—Toponimia Chimú. Fundación "Alfredo Pinillos Goicochea", Trujillo

Who Runs the Orange River Oasis? A Case Study of the Midstream Orange River Oasis, Northern Cape Province, South Africa

David Blanchon

Abstract This chapter examines the conflicts among water users in the particular setting of a fluvial oasis, the largest in South Africa, which stretches along the Orange River for almost 300 km between the Karoo and Kalahari deserts, from Boegoeberg Dam to Augrabies Falls. Since the end of the nineteenth century, the White, predominantly Afrikaner farmers who colonised this area, dispossessing the Coloured farmers and then using their workforce, levelled the alluvial sandy dunes, built canals and planted crops in less than 60 years, from the 1880s to the 1950s. Although controlling water was a shared objective among Afrikaner farmers, English businessmen and colonial hydraulic engineers, the question of who should control the water and for what purpose raised endless conflicts in colonial society. Following the completion of the Orange-Fish Inter-basin transfer (1977) and the Lesotho Highland Water Project (1998–present), the central government has the possibility of diverting a large amount of the Orange water flow outside the basin. Today, there is a growing fear in the lower reaches that if severe drought affects the Gauteng area, the decision will be made to divert all the water towards the political and economic heart of South Africa. Using an approach based on political ecology and environmental history, this chapter investigates the complex links between local and national institutions that are in charge of water management, in the context of an institutional setting rapidly evolving since the end of apartheid in 1994. It also examines how the perspective of water shortages impacts the economy of this oasis in the context of globalisation.

Keywords South Africa · Inter-basin transfers · Apartheid · Hydropolitics

---

D. Blanchon (✉)
University Paris-Ouest-Nanterre-La Défense, UMR CNRS LAVUE, Nanterre, France
e-mail: david.blanchon@gmail.com

E. Lavie and A. Marshall (eds.), *Oases and Globalisation*,
Springer Geography, DOI 10.1007/978-3-319-50749-1_6

1 Introduction

Water management issues have been debated at the highest State level in South Africa since the establishment of the first colonial settlements. The first colonial rulers considered the control of water resources a key element in the success of their settlements, and this effort to tame "predictably unpredictable" South African rivers was seen as a major task for all the ruling bodies in the country. As A. Turton (et al.) wrote in 2005: "*From the beginning of South Africa's history, water has played an important role in the shaping of the country, not only demographically but also politically*" (Turton et al. 2005: 15).

As a result, the South African waterscape is today one of the most transformed in the world, only to be compared with the American West. South African authors could write correctly that: "*water is flowing [in this country] towards money and power.*" (Basson 1997). Although controlling water was a shared objective among Afrikaner farmers, English businessmen and colonial hydraulic engineers, the question of who should control the water and for what purpose raised endless conflicts in colonial society.

This chapter will examine these conflicts in the particular setting of the fluvial oasis, which stretches along the Orange River for almost 300 km between the Karoo and Kalahari deserts, from Boegoeberg Dam to Augrabies Falls. It is the largest fluvial oasis in Southern Africa and has been developed since the end of the nineteenth century. The White, predominantly Afrikaner farmers who colonised this area, dispossessing the Coloured farmers and then using their workforce, levelled the alluvial sandy dunes, built canals and planted crops in less than 60 years, from the 1880s to the 1950s.

However, being located more than 800 km from Cape Town and Pretoria, far from any important trade route, the farmers, close to bankruptcy, very soon asked for external financial and material support. They gained it quite easily, due to the strategic location of the oasis, not far from the Namibian border, and the importance of farmers' associations in the newly formed Union of South Africa in 1910. Nevertheless, since the beginning, gaining funds has meant losing part of the control over water.

This loss of control, albeit marginal until the first Government Water Schemes (GWS) were built in the 1960s, began to be problematic when the huge dams of Bloemhof, Gariep and Van der Kloof were built in the 1970s far upstream. Following the completion of the Orange-Fish inter-basin transfer (1977) and the Lesotho Highland Water Project (1998–present), the central government has the possibility of diverting a large amount of the Orange water flow outside the basin. Today, there is a growing fear in the lower reaches that if severe drought affects the Gauteng area, the decision will be made to divert all the water towards the political and economic heart of South Africa.

Using an approach based on political ecology and environmental history, this chapter will investigate the complex links between local and national institutions that are in charge of water management, in the context of an institutional setting rapidly evolving since the end of apartheid in 1994. It will also examine how the perspective of water shortages impacts the economy of this oasis in the context of globalisation.

2 A "Nile" in South Africa?

Between Boegoeberg[1] Dam and Augrabies Falls, the Orange River[2] stretches for almost 300 km between the Karoo and Kalahari deserts (Fig. 1). In this region, summers are extremely hot (the average maximum daily temperature reaches 35 °C in January) but frost occurs during winter. At Upington, the rainfall is 190 mm and at Kakamas 130 mm, while the mean annual evaporation reaches around 2667 mm (Swanevelder 1981). This long oasis, populated by around 150,000 inhabitants,[3] is therefore totally dependent on the water of the Orange River.

2.1 A Long Fluvial Oasis

The Orange River basin covers an area of approximately[4] 1,000,000 km^2 with a mean "natural" annual run-off of 11.2 km^3. Its sources are located in the small landlocked kingdom of Lesotho and it runs first south-west and then north-west before its confluence with the Vaal. This river is often considered in South Africa as being a river basin in its own right (Turton et al. 2005: 93). The river course between the Vaal confluence and Augrabies Falls is divided into two distinct parts.[5] In the upper stretches, irrigated agriculture is not widely developed, whereas after Boegoeberg Dam, a large and continuous fluvial oasis has been built between the Karoo and Kalahari deserts.

This area could also be split into two rather different parts: from Boegoeberg Dam to a few miles upstream of Upington, the river valley is quite narrow (2–5 km), with a rather continuous and gentle slope, but from Upington to Augrabies Falls, the character of the valley changes rapidly. The river acquires a braided habit favouring the accumulation of silt to the lee of the more resistant rocks, and depositing

[1]Also known as Buchuberg Dam.

[2]The name Orange River was given in 1777 by Colonel Robert Jacob Gordon after the House of Orange, kings of the Netherlands. It was known as Gariep (Great) River by the Khoi-Khoi. In Lesotho, it is known as Senqu River.

[3]The oasis is now divided into two local municipalities: Kai Garib, around Kakamas (pop. 65,869); and Khara Hai, around Upington (pop. 93,494).

[4]As large parts of the basin are endoreic, its area varies according to different authors.

[5]In the various reports of the Department of Water Affairs, the whole area between the Orange/Vaal confluence and the mouth is labelled as the "Lower Orange".

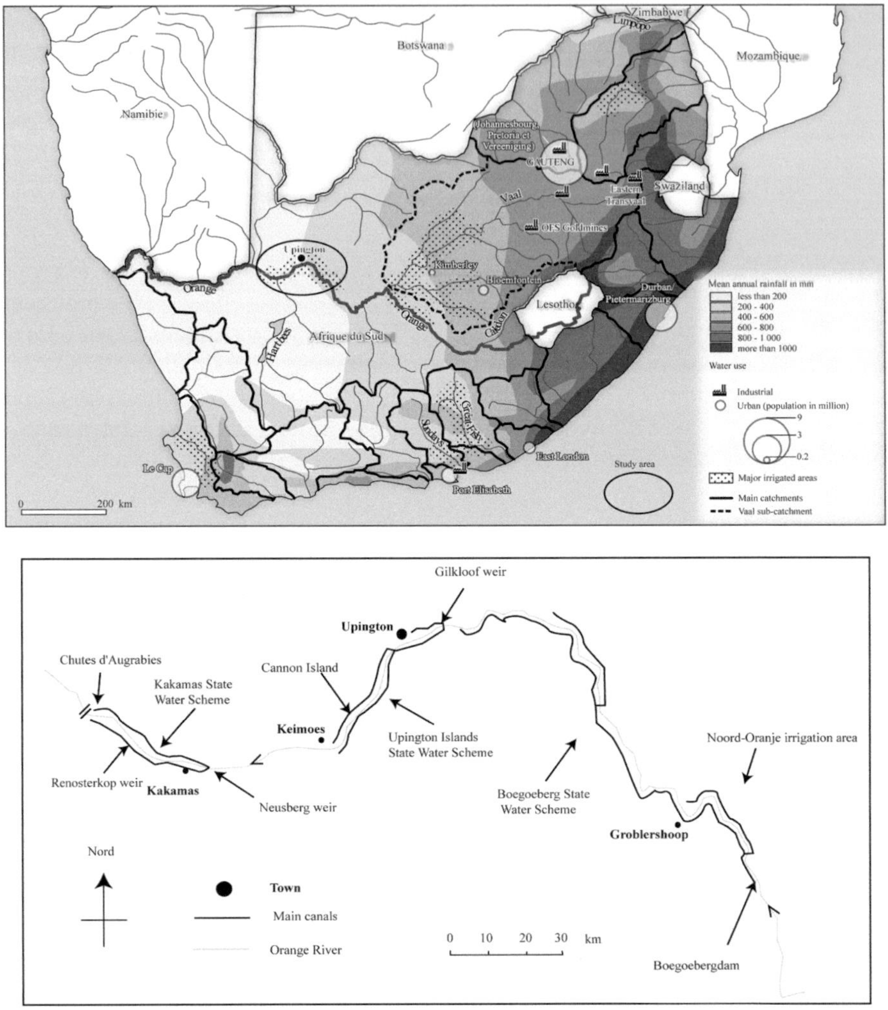

Fig. 1 Orange River Basin and close-up of the Boegoeberg/Augrabies stretch

numerous elongated islands, such as Kanon Island, 10 km long and 3 km wide (Swanevelder 1981: 35).

The downstream "natural" limit of this oasis is formed by the Augrabies Falls, where the river plunges 60 m over an almost sheer drop and then passes through a long gorge.[6] Shortly after the Augrabies Falls, the Orange River forms the border between South Africa and Namibia for almost 400 km, cutting a deep valley, until it reaches its mouth in the Atlantic Ocean.

[6]The total drop is around 160 m.

2.2 A "Predictably Unpredictable" Allogenic River

In its middle reaches, from a hydrological point of view, the Orange River has two main characteristics. Firstly, its allogeny: although only 4% of the basin area lies in Lesotho, 41% of the mean annual run-off (MAR) comes from the Kingdom, and 46% from the Vaal and the Caledon. The oasis is, therefore, totally dependent firstly on Lesotho and secondly on powerful water users in the Vaal basin, such as the Johannesburg metropolis and the industries of the Gauteng region, the irrigators in the Vaal Basin, and users who depend on the water transfer from the Upper Orange to the Fish River Valley and Port Elisabeth.

The second characteristic of the Orange River is the high natural variability of the annual run-off. Like many other South African streams, the flow of the Orange River is "predictably unpredictable" (Davies and Day 1998). At Boegoeberg, the "natural"[7] Mean Annual Runoff (MAR) is 327 $m^3\ s^{-1}$ or 10.3 km^3 (DWA 1992), with high flows during the summer months (January to March) and low flows during winter. However, depending on the rainfall in the Lesotho Highlands, large variations have been monitored. Before the construction of big dams, the river often almost ceased to flow during the winter months and during the hydrologic year 1932–1933, the total run-off was only 1.3 km^3 at Boegoeberg.

On the contrary, the occurrence of destructive floods is well acknowledged. Flows greater than 8000 $m^3\ s^{-1}$ were measured in 1934, 1944, 1950, 1958, 1967, 1974 and 1988. The maximum flow was measured in 1925 in Boegoeberg at 16,200 $m^3\ s^{-1}$ in March, before the construction of big dams. Nevertheless, a flood of 5200 $m^3\ s^{-1}$ occurred in January 2011, causing heavy damage to the irrigation infrastructures, one year after the 2010 flood. According to Hoffie Joubert, chairperson of Agri Northern Cape's Irrigation and Water Affairs, "*the total flood damage in Northern Cape amounts to R932 million* […] *About 800 commercial farmers in the Lower Orange were hit by the floods, of which nearly 30% may not be able to continue farming. More than 5,000 permanent and seasonal farm workers are affected, and food security is also at risk*" (quoted in *Farmer's Weekly*).[8]

2.3 A "Nile" Too Far from South Africa's Economic Heart?

These two combined features, allogeny and variability, led the first explorers to describe the Orange River as the "Nile" of South Africa. In 1877, J.C. Brown wrote: "*From periodically overflowing its banks, and depositing alluvial karoo-mud, the best of manures, this river has been called the Nile of South Africa, and we*

[7]The "actual" MAR is thought to be around 5.5 km^3, due to diversions (see below for details).

[8]24 February 2012: "The Lower Orange River was hit by floods twice in a row in 2010 and 2011" by *Annelie Coleman.*

question whether anywhere in Egypt the husbandman reaps more than a hundredfold from his seed, which is here common" (Brown 1877: 260). Sixty years later, C. Birkby used almost the same words, declaring that: "*this river, which the natives and the older Boer folk still call the Gariep, is the Nile of South Africa.*" *Periodically it overflows its banks and like the Nile leaves a dressing of rich silt that forms soil of amazing fertility*" (Birkby 1936: 222).

Yet, it is a poor, distant Nile, far from being the heart of the country, and reports from irrigation engineers, since the end of the nineteenth century, have been much less enthusiastic. F.E. Kanthack wrote in 1917 about the Orange River: "*From the point of view of practical irrigation, however, it has a most unsatisfactory flow. Like so many of our South African rivers, it fails just at the time when water is most urgently required, in the spring and early summer months of August, September and October, and the flow even at the best times, in these months, is quite insufficient to meet the requirements of even the little bit of irrigation at present established*" (U.G. 29-17: 35).

In fact, the history of the Orange River fluvial oasis could be read through the tensions between the "dream" of a cornucopian valley in the desert and the harsh reality of agricultural development on alluvial soil plagued with drought and floods, and far from any important market. Upington is located more than 800 km from Johannesburg and Cape Town, via long and poorly maintained roads crossing hostile deserts.

3 A History of Violent and Conflictive Settlements

Located at the northernmost periphery of Cape Colony, the Orange River was not reached by European voyagers before the end of the eighteenth century, almost 120 years after the foundation of the city of Cape Town in 1652. The first White settlements were not established before the end of the nineteenth century.

Little is known about life in the valley before the arrival of the first Europeans. Wilcox (1986: 20) gave a picture that may not be far from reality: "*here was semi-desert to the north and to the south, with the Great River the only perennial river for some 300 km and with but a few springs, so grazing cattle was only seasonal, after the scanty rains. For the Hottentots, therefore, not agriculturists and only occasional hunters and fishers, access to the Orange River Valley for much of the year was practically vital. The competition was made yet fiercer by the migration of other Hottentot tribes ... from the southern Cape where they were being ousted by the European farmers. The increasing population of Hottentots and their livestock drove out the wild game on which depended the Bushmen, who had prior occupation of the valley.*"

3.1 A "Lawless Frontier"

Due to population displacements in the rest of the colony, according to Penn (1995) "*by the 1780s, the river had become a zone of terror where frontiersmen of varied origins—Europeans, slaves, "Bastaards," Khoi, "Bastaards-hottentots, San and Oorlams*[9]*—formed predatory commando gangs which robbed weaker groups of their livestock*" (Penn 1995). Moreover, for almost a century, as Wilcox (1986: 54) wrote: "*The region of the Orange River at this time has a claim to have been the most lawless frontier in the world compared to which the American "wild west" was highly civilised.*" This also adds weight to the comparison between South African and American waterscapes.

In 1847, the Orange River formed the frontier of the British Cape Colony with Griqualand to the east and British Bechuanaland to the west. However, the frontier was far from being controlled, and several military expeditions were sent in order to pacify the zone and stop the raids that were made towards the southern parts of the colony. Sir Water Currie, who commanded the Colony forces during the two so-called Kora wars (or Koranna wars), reported about the Upington Islands that "*I used to think that the Fish River was a stronghold, but it stands nowhere in comparison with this water jungle.*" According to Ross (1975: 562), "*until they were finally crushed in 1879, the Kora formed the most significant non-"bantu" adversaries of the expansion of the colony.*"

It was not until the end of the nineteenth century that the frontier of Cape Colony was pushed further north, where it still stands, and that the region of Upington became "pacified" from the colonial point of view.

3.2 The Hidden Face of Irrigation

When the frontier was fixed in the newly formed magisterial district of Gordonia[10] that covered most of the fluvial oasis from Boegoeberg to Augrabies Falls, "*the inhabitants of Gordonia were mainly Baster (a term of flexible applicability), with a few whites at first largely related to them by marriage, as well as remnants of Kora, San and some Xhosa*" (Legassick 1999: 374). Even at the beginning of the twentieth century, in the 1904 census, 1712 Europeans, 2370 "Hottentots," 1245 "Kafir and Bechuana," and 3888 "Mixed and Others" were counted.

In a small museum in Upington, at the beginning of the twenty-first century, the history of irrigation was presented as follows: in the 1870s, Reverend Schröder arrived in the underdeveloped region and dug the first furrow in the 1880s in order

---

[9]According to Penn (1995), Bastaard—white-Khoi, or white-slave; Oorlam—Khoi who had acquired some cultural colonial trappings.

[10]It was named after Sir Gordon Sprigg, who was prime minister of Cape Colony four times between 1878 and 1902.

to irrigate the islands in the river. Later, Boer pioneers settled in the region to make it one of the most productive agricultural regions in the country. Further downstream, the story was almost the same, with the foundation of the Kakamas Labour Colony in 1897 by B.J.P. Marchand for the then so-called "Poor Whites."

Recent research has described a quite different story. According to Legassick (1999), most of the land was already cultivated by "Baster" farmers (latterly considered Coloureds), such as the Abraham September family, in the 1870s. Some irrigation works were already in place when Reverend Schröder arrived. The "Basters" were then progressively ousted from their land by Boer farmers. In 1882, 81 farms were allocated to Basters; in 1893, there were only 33 against 44 White farms while in 1920, only 6 farms were still owned by Basters.

The dispossession of Basters by mostly Afrikaner farmers is acknowledged by administrative reports. For instance, E.A. Nobbs wrote in 1903: "*Mr Schroeder [...] worked here among the Korannas, and later it was a flourishing Bastards settlement, now the land is mostly in the hand of Boers.[...] the number of Bastards holding land is steadily diminishing*" (Cape of Good Hope 1904a: 40).

The story of irrigation is still subject to debate and conflicting memories in the region. C.M. Gouws reported that during her visits to Upington in 2005: "*For the white irrigation farmers, the legacy of Reverend Schroder as the "father" of Upington and the innovator of the canal is clearly important. [...] More than once, when respondents were interviewed and asked for comments on the matter, their reactions had a racist undertone. One of the farmers said bluntly that Abraham September's legacy is nothing more than digging a little furrow to direct sewage away from his house to the river.*" (Gouws 2012: 143).

3.3 A Stronghold of Apartheid

During the segregation (1910–1948) and apartheid (1948–1994) eras, the oasis was one of the regions where racial policies were strictly enforced. Coloured property, under the Group Area Act,[11] was restricted to a small reservation at Eksteenkuil, upstream of Keimoes, and most of the Coloured population, who formed the vast majority of the workforce, were employed on White farms in very poor conditions. According to Reverend Abrahamson, quoted in Robbins (1986: 156), in the 1980s: "*Whites have swindled the Coloureds out of a lot of their land. Most of it, in fact. And the Group Area Act hasn't helped us at all. I know a young fellow who can no longer inherit his father's freehold land because it has been proclaimed White. If you want to see apartheid, go to the drive-in, either here or in Keimoes, and look at the wall built down the middle.*"

[11]The Group Area Act was passed by the South African parliament in 1950. It assigned racial groups to specific areas and put a strong control on the acquisition and possession of property by different racial groups. This Act gave a legal basis for forcibly removing the non-Whites living in "White" areas.

On the contrary, White farmers benefited from State subsidies, not only for agriculture but also for educational grants, improved access to urban services and support for agricultural cooperatives.[12]

It was the period of the most impressive development of the oasis. From a few hectares at the beginning of the twentieth century, the Beplanningkomittee vir Benede-Oranjerivier (1973) listed 38 major irrigation schemes in 1973, with 22,630 ha irrigated. In less than 50 years, almost all the islands in the river between Boegoeberg and Augrabies were transformed into highly productive commercial farms, all owned by White, mostly Afrikaner, farmers. Gordonia, in the middle of the apartheid era, was typical of South African society at that time: the Whites were less than 25% of the total population but owned 100% of the land and almost all the economic power. Moreover, the percentage of Whites was slowly decreasing, from almost 42% in 1921 to 25% in 1961 and less than 10% now.[13]

This brief historical overview of the Upington area shows that lawlessness, dispossession, segregation and apartheid were its main features: the history of the oasis is far from peaceful. This historical background explains the composition of the population dominated by a vast majority of Coloureds, of mixed descent, who were dispossessed, exploited and politically marginalised until less than 25 years ago. This is why the region is still an illustration of a highly divided society, between White farmers, still owning much of the land and wealth of the region, and Coloured workers. Increasing unemployment and crime rates, higher than the South African average,[14] are the direct heritage of this violent past.

4 "Of Land and Water," Government and Farmers: Who Is in Control?

The history of water management reflects the profoundly unequal national policies of segregation. Due to the special features of the hydrology of the Orange River, the oasis waterscape is the result of a complex nexus of competition between local farmers, irrigation engineers and the central government. Twenty years have passed since the end of apartheid but little has changed, notably due to the inertia of

[12]For instance, the Orange River Wine Cellars Co-op Ltd. was founded on 23 December 1965 in order to help the farmers produce wine.

[13]The population in Gordonia Magisterial district in 1921 was 7800 Whites, 9500 Coloureds and 980 Bantus; in 1960 it was 17,960 Whites, 44,400 Coloureds and 10,314 Bantus. In the municipality of Khara Khais (the boundaries are different), the racial makeup in 2011 was 23.1% Black Africans, 65.2% Coloureds and 9.9% Whites.

[14]Which is already very high at 31/100,000 (world average 6/100,000).

hydraulic infrastructures. In any attempt to understand what is happening in this oasis today, the impact of "antecedent political systems" must be fully considered.

4.1 "Land Pirates" and Water Wars

As mentioned above, the Orange River was a loosely controlled frontier and, for a long period, it was unclear to whom land, water rights and power were supposed to belong.

It was generally assumed that most of the land in the valley was crown or government property. F.E. Kanthack assessed in 1917 that "*along the Orange River, below Prieska, the Government owns an immense amount of land on both banks and also many islands in the river*" (Union of South Africa 1917: 33). However, in 1904, a report to the director of irrigation stated that "*uncertainty exists regarding the ownership of the islands. Most of them appear to be undoubtedly crown land, but some at least are claimed by the owner of adjoining farms*." (Cape of Good Hope 1904b: 7).

Consequently, these documents suggest that most of the land was settled without legal property rights. For example, in the late 1920s, poor White farmers began to clear and strip an island downstream of Upington, which is now known as Cannon Island, to expand cropland. They were firstly considered "land pirates," but eventually these illegally acquired properties were secured in 1936 by the Crown Land Act.

Like land properties, water rights were also plagued with uncertainty. F.E. Kanthack described in 1917 the local "water wars" that were directly linked to the "predictable unpredictability" of the river flow. "*In a dry year,* he wrote, *a state of anarchy may be said to prevail along the river. Weirs and dams are breached by lower owners, generally with the aid of explosives*" *[…] Farmers suddenly find that water which should naturally flow down to them has been diverted. The injured farmer promptly retaliates by blasting a reef somewhere else, and it goes on until a lucky flood puts an end to these illegal practices.*" *[…] During times of very low water, irrigators have been known to claim their rights to water and defend their unlawful acts with the aid of a rifle*" (Union of South Africa 1917: 33).

However, such a situation was considered unacceptable for an irrigation engineer like F.E. Kanthack; therefore, he advocated "*a complete [governmental] control of the water in the river at reasonable intervals, so that an equitable distribution of water can be effected in each section of the river controlled,*" saying in the same report that "*the irrigation works on the islands are very primitive*" (Union of South Africa 1917: 33).

The Orange River irrigation history illustrates the long-term fight to control land and water between irrigation engineers and farmers, Irrigation Boards and the Department of Irrigation (created in 1910, the Board became the Department of Water Affairs after 1956), with tensions rising from below, such as the difficult balance between subsidies and control.

4.2 Subsidies and Control

Even when the irrigation rights were partially settled through the establishment of Irrigation Boards, there were unresolved problems, such as funding and the risk of bankruptcy, which led farmers to ask for government subsidies. In exchange, the Department of Irrigation negotiated the entire control of the Boards. However, the farmers, through the Riparian Rights,[15] their influence on the local offices of the Department of Irrigation and their political power at the national level, managed to secure a large control over water, in spite of government investments in the construction of hydraulic infrastructures. Most of the time, the central government took control of the water infrastructures for a few years only, and then handed over their management to the local Irrigation Boards.

The following analysis of three case studies, namely Kakamas, Upington and Boegoeberg, which cover most of the fluvial oasis, shows the complex history of irrigation institutions.

At the lower end of the oasis, a few kilometres above Augrabies Falls, irrigation began with the foundation of the Kakamas Labour Colony for Poor Whites in 1897. However, despite the involvement of the colonists, the Labour Colony soon asked for support.[16] Following this 100% subsidy, the Kakamas management board was still empowered to "*control and maintain irrigation work and to distribute water therefrom until 1964, when the Kakamas irrigation works were handed over to the State by the Supervision and Control and Authorised Agent of the Dutch Reformed Church*" (ibid). The reason given was that: "*All the irrigation schemes in the Republic as well as abroad are subsidised to a greater or lesser extent on the grounds that not only the irrigators but all the sections of the community derive benefit from the undertaking and it is therefore not proposed that the irrigators under the Kakamas Government Scheme should repay the full cost of the betterment works*" (ibid: 4).

A similar pattern was observed in the Upington area, where several Irrigation Boards were established in the first decades of the twentieth century, such as Straussburg in 1925 and Louisvale in 1918, in order to settle disputes about water rights. In the latter, for instance the farmers were unable to repay their loans so, as in Kakamas, they asked for government help in exchange for losing part of their control of water. According to a 1966 report: "*in 1921 an irrigation loan of 8,000 rand was granted […] by 1936 the amount due, including interest, amounted to R. 9,000. The Parliamentary Select Committee, however, agreed to write off the*

[15]Reaffirmed in the Irrigation Act of 1912.

[16]"Subsequent to the establishment of the Union in 1910, State financial aid, in the form of loans, was granted to the Settlement for the construction of canal and betterment works and a total amount of R 95,250 was made available. In 1950, a subsidy of 50% was granted by the State in respect of the cost of this work […] It soon became evident that the Management Board would be able to achieve little unless a higher subsidy was made available to it […] The Program of R. 500,000 was thereafter proceeded with on the basis of a State subsidy for the full amount (for 1955–1956), then 850,000 in 59–60, with 100% subsidy" (Republic of South Africa 1965: 2).

amount [...] The Ministry of Water Affairs administered the scheme during the construction of the new canal in 1936, but the Board resumed full control in 1941" (Republic of South Africa 1966: 4).

The story of Boegoeberg is slightly different because, in this area, the government was involved in the construction of this scheme from the beginning. The Department of Irrigation was very reluctant to build a dam in Boegoeberg, and motions brought before parliament regarding the irrigation of this area were rejected in 1895 (after reports made in the 1890s) on the grounds that it would be too expensive. The Boegoeberg scheme was also discarded in a 1917 report and in A.D. Lewis' report for the years 1928–1929, stating that this scheme "*is likely to be exceedingly expensive on account of the great length of furrow compared with the limited extent of irrigable land and the necessity for storage provision*" (Union of South Africa 1930: 36). The only reason why the Boegoeberg dam was finally built is the world economic crisis of 1929, which contributed to the enrolment of unemployed workers for the construction of hydraulic infrastructures, as occurred in the US. According to Turton et al. (2005: 164): "*The Director of Irrigation received "sudden instructions," in 1930 to organise and start with the construction of irrigation works at Buchuberg, Oukloof and Kalkspruit. Regarding the Buchuberg scheme, his instructions were, to start "construction as soon as possible to provide employment for white people.*"

Three elements can be derived from these examples that still shape the midstream Orange River oasis waterscape.[17] Firstly, the situation of the oasis, which is located far from major markets, has been very fragile and has remained disconnected from South African economic centres. The variability of the river, with dry years and destructive floods, has posed a constant threat to the farmers' economic status, and they have never been in a position to be resilient without external help. Therefore, although the farmers presented themselves as pioneers in a harsh environment, the oasis was always dependent on the State. In 1973, for instance, only 8223 irrigated ha belonged to "independent" Irrigation Boards while 10,319 ha were included in GWS.

Secondly, the Department of Irrigation (1910–1956) and then the Department of Water Affairs (1956–1994) have been reluctant to invest large amounts of money in this remote area. In 1909, F.E. Kanthack (Director of the Department of Irrigation from 1910 to 1920) wrote that "*Every Island in the river is not irrigable. [...] Any large comprehensive scheme for the irrigation of vast areas is, I think, out of the question. Markets are very limited now, and the very costly land reclamation works are only possible now on account of high level yields and very big prices*" (Cape of Good Hope 1910). In 1929, his successor at the head of the Department from 1921 to 1941, A.D. Lewis, stated: "*My chief and great regret in presenting this summary is that I cannot put forward for the Orange a single scheme of any magnitude and of such merit as to justify the great optimism which at present prevails*" (Union of

[17]For a comprehensive view of the Orange River Basin irrigation history, see Turton et al. (2005: 88–262).

South Africa 1930: 24) and a few pages later he added: "*We have got to face that South Africa will never be a great irrigating country*" (ibid: 30).

Thirdly, interactions between farmers and water engineers in this region have been fuelled by continuously strained relations, which were not improved with the implementation of large inter-basin transfer (IBT) schemes in the 1960s. This, in turn, exacerbated the tensions that still prevail today.

4.3 The IBT in South Africa: A River No Longer?

This loss of control by the farmers, albeit marginal when the State-controlled works started in the 1920s and even when the State decided to build the Boegoeberg dam in the 1930s, began to be problematic when the large dams of Bloemhof, Gariep and Van der Kloof were constructed in the 1970s far upstream. Following the completion of the Orange-Fish inter-basin transfer (1977) and the Lesotho Highland Water Project (1998–present), the central government has the possibility to divert a large amount of the Orange water flow outside the basin (Fig. 2). There is a growing fear in the lower reaches that if a severe drought were to affect the Gauteng area, all available water would be diverted towards the political and economic heart of South Africa, that is the Gauteng.

The fears of the farmers of the Upington area were aroused by declarations, notably those of preeminent scientists such as D.C. Midgley, who wrote in the *South African Journal of Sciences* at an opening conference about the Orange River Development Project: "*the primary objective of the ORDP is to transform this*

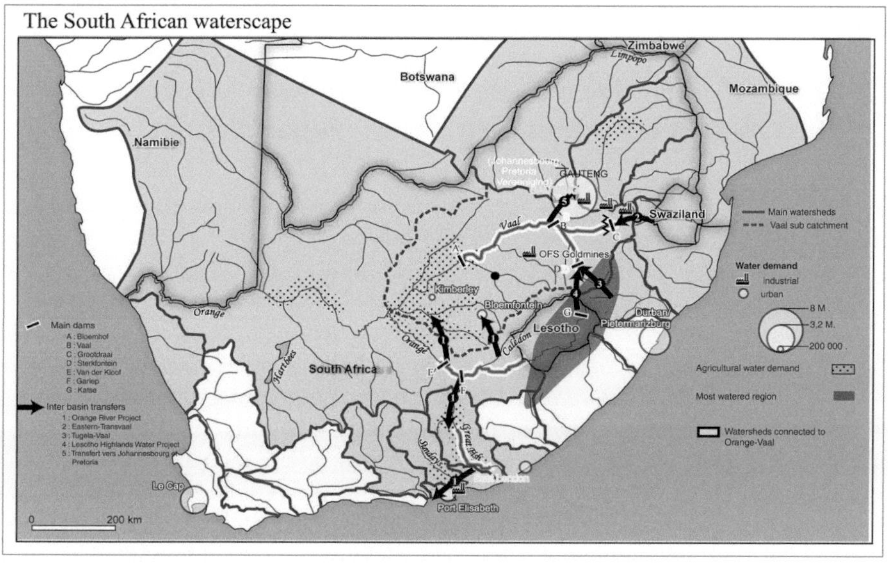

Fig. 2 Inter-basin transfer (IBT) in South Africa

intermittent isolated river [...] into a perennial distributary system, designed to spread the benefits of a controlled water supply far into the Karoo and into the Kalahari [...] If we are to survive, the former pattern of small plot irrigation *settlements must surely give way to large, highly organised scientifically operated farms*" (Midgley 1963: 461–462).

His predictions have almost entirely come true, with the flow of the Orange River profoundly altered on the one hand, and placed under the control of the Department of Water Affairs, on the other. As a result, the Vaal River has been completely disconnected from the Orange River. About 10% of the total "natural" MAR, which is more likely to be 20% of the actual run-off, if the Vaal is no longer considered a tributary of the Orange, could be diverted towards the Fish River out of the basin at Gariep Dam (known as H. Verwoerd dam when it was constructed). In the case of low rainfall on the Lesotho Highlands, almost 80% of the flow could be diverted. In this perspective, the Upington area may be considered the "sink" of the Vaal system, which is still centrally controlled by the DWA (Bourblanc and Blanchon 2014).

As predicted by Midgley at the local scale, all the Irrigation Boards were put under the control of the State, with the implementation of GWS covering the Upington area and Kakamas in the 1960s in exchange for large investments. As stated in different White Papers, the schemes were to be directly controlled and administered by the State. Although the farmers still controlled the land and retained a formal control over water allocation, by the end of the 1980s, they had lost most of the actual control of water flows to the DWA. By this time, the DWA, alongside the State and Parliament, was still almost entirely managed by people "that could be trusted," mostly Afrikaners. However, since 1994, the situation has evolved towards the inclusion in government institutions of a previously excluded black majority.

5 Losing Control Over Water, Opening New Markets: The Oasis in the Globalised "New" South Africa

Along the Orange River, as in the American West, for most of the twentieth century, two roughly equivalent centres of power competed: "*a private sector of agriculturists and a public sector made up of bureaucratic planners and elected representatives*" (Worster 1985: 51). However, as D. Worster noted "*Neither group is autonomous. Both need each other, reinforce each other's values, compete for the upper hand without lasting success, and finally agree to work together to achieve a control over nature that is unprecedentedly thorough*" (ibid).

However, with the end of apartheid, the fragile Orange River hydraulic arrangements collapsed. In the "new" South Africa, where the vast majority of the population has been dispossessed and deprived of the profits derived from water infrastructures, the water control apparatus (and its chief profiteers, the White farmers) has been criticised, and a new water law was designed in order to change radically the water management system.

Meanwhile, the economic situation of the country has changed dramatically, particularly for White commercial farmers: new markets have emerged with the end of the economic boycott, while farming subsidies have almost completely disappeared.

As a result, these two "revolutions" have profoundly affected the Orange River oasis waterscape.

5.1 Losing Control Over Water? from the Texts to their Implementation

As far as the water sector is concerned, a new water act was passed in 1998 (Act 36 of 1998) replacing the 1956 one.

This new act proposed a complete reframing of South African water policy, with a strong political commitment to erase the injustice of the apartheid era.[18]

Two new institutions were defined:

- at the basin level, Catchment Management Agencies (CMA) would have to "investigate and advise on the protection, use, development and control over water in the catchment, and develop a catchment management strategy";
- at the local level, water would be managed by newly formed Water Users Associations (WUA), which were supposed to replace the Irrigation Boards.

The CMAs were also supposed to be the symbol of the post-apartheid water policy, shifting the power from the highly centralised and powerful Department of Water Affairs in Pretoria to the local communities, notably "disadvantaged persons or communities which have been prejudiced by past racial and gender discrimination in relation to access to water" (RSA 1998: 90). According to the DWAF, "*the purpose of establishing these agencies is to delegate water resource management to the regional or catchment level and to involve local communities*" (RSA 1998: 85). Besides, the existing water rights would be transformed into water licenses, which would be delivered by the DWAF.

This new water law aimed to open up a new era for the oasis, with a local control of water resources and a shift of power from White farmers to a larger community of water users.

[18]In the introduction of the White Paper on National Water Policy, it is stated that: "The colonial law-makers tried to use the rules of the well-watered colonising countries of Europe in the dry and variable climate of Southern Africa. They harnessed the law, and the water, in the interests of a dominant class and group which had privileged access to land and economic power.

It is for this reason that the new government has been confronted with a situation in which not only have the majority of South Africa's people been excluded from the land but they have been denied either direct access to water for productive use or access to the benefits from the use of the nation's water" (RSA 1997).

As for the implementation of CMAs, in a press release in March 2012, the South African Minister of Water Affairs stated: "*The Minister decided to reduce the number of CMAs to nine from the original proposal of 19 CMAs. This is due to a number of reasons including the technical capacity required to staff CMAs, and the challenges such a large number of institutions poses to the Department of Water Affairs (DWA) in regulating their performance*." The above-noted necessity to reduce the number of CMAs is not particular to the Lower Orange CMA, which was supposed to cover the basin from the Orange–Vaal confluence to the mouth, but demonstrates that there will be no locally defined water strategy in the near future.

Whereas the recent literature provides several comments on the difficulties encountered by the CMAs and the water licensing process, at the local level the Water User Associations appear to be well-functioning institutions. Three of them were created along the oasis, at Boegoeberg (covering most of the ex-Boegoeberg GWS), Upington and Kakamas. As a witness of these WUAs in 2000, I noticed that they were clearly dominated by White farmers in the region: at the Kakamas WUA Steering Committee, 8 of the 15 members were White farmers, with only one representative of "emerging farmers" (i.e. Coloured) and one representative from the DWA (Blanchon 2003). A few years later, the same situation was acknowledged in the Upington Islands WUA, which "*is monopolised by irrigation farmers and the irrigation boards that served them in the past. In contrast to the (developing) LOCMA, the UIWUA seems to be functioning remarkably well. In many respects its operations are even better than those undertaken by the Khara Hais Local Municipality. This local authority is in disarray*" (Gouws: 169).

In this area, WUAs are the only efficient water institution, which in turn raises serious doubts about how little the water control has changed, 17 years after the passing of the new water law. Furthermore, there is evidence that White farmers have "regained" control over DWA engineers, in particular at the local level through their appropriation of WUAs. This is also a consequence of the loss of resources by DWA local offices (capacity and funding) since 1994.

5.2 Entering the World Markets

The end of apartheid has also introduced changes into the economic situation of the oasis with, on the one hand, the end of farm subsidies and, on the other hand, the opening of markets and thus new investment opportunities.

At the beginning of the 1990s, the most profitable crop was raisins for the local market, which were cultivated on the alluvial islands of the Orange River. However, as soon as the economic sanctions were lifted, it appeared that this region was particularly suitable for the cultivation of table grapes, especially for European (and secondly North American) customers. Due to its position in the Southern Hemisphere and its huge insolation, table grapes can be harvested in December, and thus be ready for the Christmas holidays, when the demand (and prices) soars.

Nevertheless, the shift from raisins to table grapes demands a huge investment and is very risky. The prices can drop very quickly after Christmas, and the production of grapes varies greatly from year to year. Only a few White commercial farmers could make the transition, and move from a labour-intensive farming to a highly capital-intensive commercial agriculture. One example is Karsten Farms, which is located not far from Upington and has 866 ha under irrigation, using an average amount 10,000 mm of water per hectare per year (Gouws 2012: 148). Most of the White farmers, whose properties seldom reach 30 ha, could not afford to make such investments, and there is a growing gap between large and small commercial farmers.

These processes have led to one of the most remarkable changes in this oasis waterscape: the "outer land" (Fig. 3), located outside the alluvial islands, which was

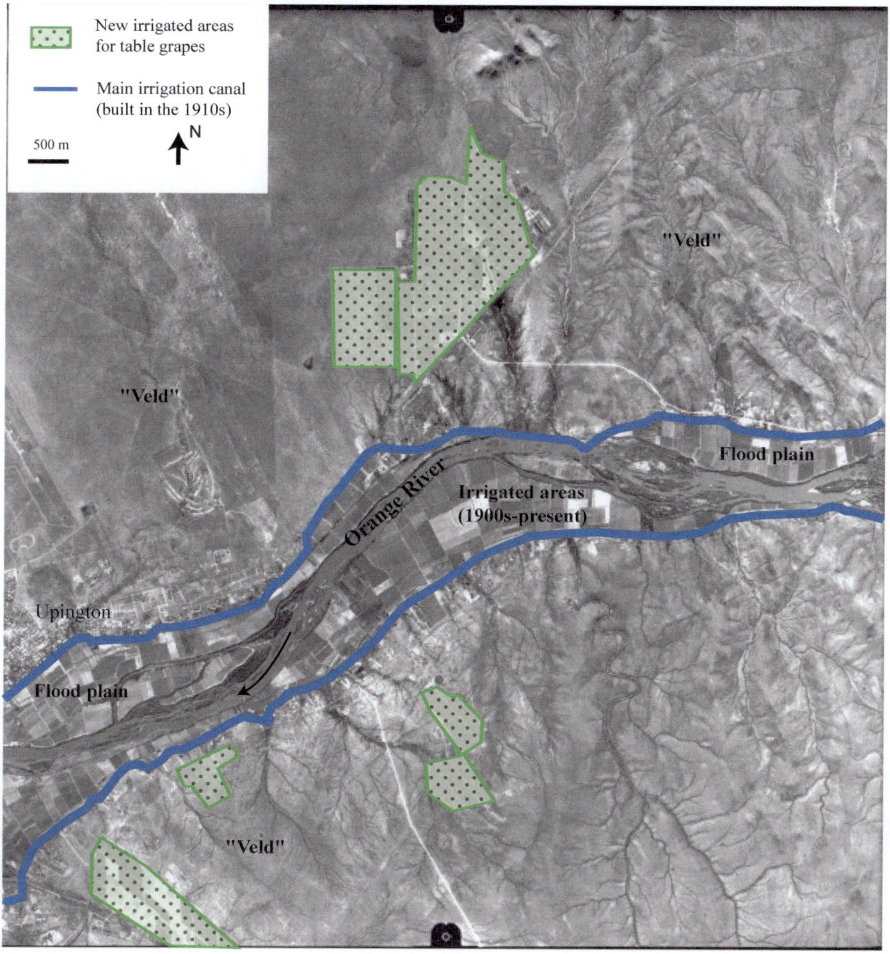

Fig. 3 New development on the "outer land"

used by pastoralists and never before irrigated, has now become the site of the most profitable South African agricultural development. This change was linked with the establishment of a local water market in the late 1990s (Armitage 1999). Irrigation farmers who do not use their water rights sell them at an average price of 5000 to 15,000 m^3/ha to table grape growers in the "outer land" (Armitage et al. 1999).

5.3 The Fate of the Oasis

With the development of the "outer land," it seems that the Orange River oasis has largely taken advantage of the end of apartheid, and that it is well placed in a globalised economy. However, the recurrent problems of the oasis—climatic variability, lack of control over its water resources and remoteness—have not disappeared.

Climatic variability poses a serious threat to the economic viability of the table grape industry: early summer rains in the first ten days of December can delay the harvest until after Christmas and generate huge losses for farmers, as the prices drop by almost 50% in January.

Moreover, the oasis is still under threat of a "water management-induced" water scarcity. The absence of a local CMA underlines that the power of water management at the national scale remains in the hands of the DWA central office. Local authorities have already complained about this situation. For instance, in 1995, the premier of Northern Cape Province, in a letter to K. Asmal, Director of the DWA, wrote that: "*the Vaal, major tributary of the Orange River System, does not contribute to flow in the Orange River any longer, except during major floods […] Although the province has jurisdiction over the biota of the Orange River […] the management of the water resource as such is a national function. The province is thus unable to control the most important physical factor affecting the integrity of the riverine environment and its ecosystems*" (Letter from Dipico to K. Asma 1995).

This situation is entirely acknowledged by the National Water Strategy, which stated: "*the Lower Orange Water Management Area is without doubt the WMA most impacted upon by upstream development. It is the most downstream of 5 CMAs covering the Orange/Vaal Basin, with extensive Inter Basin Transfers into and from most of the WMAs*" (Proposed First Edition National Water Resource Strategy August 2002, p. D14.4). As stipulated by the new water act, a reserve of water must be provided for the survival of Namibia and the Orange River estuary. Consequently, a report concluded unsurprisingly one year later that "*Apart from the development of an additional 4,000 ha irrigation for the settlement of emerging farmers, no meaningful change in the requirements for water are foreseen in this sub-area* (p. 27)" (M.S. Basson, J.D. Rossouw, National Water Resource Strategy).

In this perspective, several calculations in a DWAF report emphasise that it would be more in line with national legislation to supply more water to the Gauteng province than to the Lower Orange CMA.

	Orange River	Gauteng	Ratio
Production (Rand per cubic metre)	0.81	198	1:244
Jobs (Jobs per million cubic m)	24	1940	1:80

Source Orange River Development Project Replanning Study, Comparative Economic Impact Analysis, 1999

Since 2010, it is clear that the fluvial oasis position is deteriorating, notably as a new dam—Polihali—is under construction in the Lesotho highlands in order to transfer more water to the Gauteng. This new transfer is planned to become effective in 2018 and, in the event of a dry year, as much as 50% of the flow could be diverted to Gauteng from the Lesotho Highlands. Considering that a reserve (a minimum flow) must be retained for Namibia, virtually no water would be left for the Orange River fluvial oasis.

As the political and economic weight of the oasis is insignificant compared to Gauteng province, if a drought occurs in the Orange–Vaal River upper basin, this scenario could not be excluded, and the latest economic boom of the oasis would be put to an end.

6 Conclusion: From Climatic Variability to Man-Made Uncertainties

Twenty years after the end of apartheid, the situation of the Orange River fluvial oasis has radically changed. The old system of water management and control, which relied on weirs and canals watering alluvial soils, has been gradually replaced by large pumps, removing water directly from the river to the "outer land." This is where the economic future of the area now lies, with a highly productive use of water. Table grapes are producing "more cash per drop" than any other variety of crops in the oasis, which illustrates the "water use efficiency" policy of the DWA.

However, as the hydraulic variability of the Orange River has been progressively reduced by large dams and IBTs, new uncertainties have appeared. In a globalised economy, the oasis is more dependent on fluctuations in world fruit markets and currency exchange rates than variations in the Orange River flow. Moreover, a democratic South Africa faces "majority choices" to divert water to the most populated places rather than to a remote oasis.

References

Armitage RM (1999) An economic analysis of surface irrigation water rights transfers in selected areas of South Africa. Water Research Commission, Pretoria, 1999 (WRC report 870/1/99)

Armitage RM, Nieuwoudt WL, Backeberg GR (1999) Establishing tradable water rights: case studies of two irrigation districts in South Africa. Water SA 25(3):301–310

Basson MS (1997) Overview of Water resources availability and utilisation in South Africa. DWAF, Pretoria, p 72
Beplanningkomittee vir Benede-Oranjerivier (1973) Verslag, May 1973. Upington 152 p
Blanchon D (2003) La nouvelle politique de l'eau en Afrique du Sud: vers une gestion environnementale des ressources? L'Espace géographique 1–2003:21–29
Birkby C (1936) Thirstland Trek. Faber & Faber Ltd, London, 224 p
Bourblanc M, Blanchon D (2014) The challenges of rescaling South African water resource management: catchment management agencies and interbasin transfers. J Hydrol 519(part C):2381–2391
Brown JC (1877) Water supply of South Africa and the facilities for the storage of it. Oliver & Boyd. Tweedale Court, Edinburgh
Cape of Good Hope (1897) Special report on inspection made by Messrs AS, Le Roex MLA, Litchfield HC, AMICE, of proposed irrigation schemes on the Orange River. Government Printers, Le Cap, G 83-'97
Cape of Good Hope (1904a) Report by the director of irrigation on his tour through the North-Western Districts. Government Printers, Le Cap, G 94-1904
Cape of Good Hope (1904b) Miscellaneous reports (p. 40 et sq.) Government Printers, Le Cap, G 94-1904
Cape of Good Hope (1907) Buchuberg irrigation works, Orange River Prieska. Government Printers, Le Cap, G 2-1907
Cape of Good Hope (1910) Report of the director of irrigation for the year 1909. Government Printers, Le Cap, G 32-1910
Coleman A (2012) The lower Orange River was hit by floods twice in a row in 2010 and 2011. Farmer's Weekly, 24 Feb 2012
Davies BR, Day JA (1998) Vanishing waters. University of Cape Town, Le Cap, 487 p
Department of Water Affairs (1992) Orange River system analysis (1989–1993), Ref. DWAF WPAM 5410 à 6688. Description of the physical systems, n° 6410
Department of Water Affairs and Forestry (1989–1993) Orange River system analysis (1989–1993), internal report WPAM 5410 to 6688
Department of Water Affairs and Forestry (1996–1999) Orange River development project replanning study: 1996–1999, internal report available at Directorate: Water Resources Planning
Gouws CM (2012) Responding to water demand strategies: a case study in the Lower Orange catchment management area (LOCMA) Thesis submitted for the degree Philosophiae Doctor in Development and Management
Kanthack FE (1909) Orange River Islands—Upington to Kakamas. In: Cape of Good Hope, Report by the Director of Irrigation for the year 1909. Government Printers, Le Cap, 1910 (G 32-1910)
Legassick M (1999) The racial division of Gordonia. Kronos J Cape Hist 25:152–186
Lewis AD (1912) Report on flying reconnaissance of Lower Orange River made by AD Lewis. In: Union of South Africa, Report of the Director of Irrigation for the period 1st January 1912 to 31st March, 1913. Government Printer, Le Cap, 1945 (U.G. 2-'14)
Meissner R (2000) The case of the Orange River. Conflict Trends 2(2000):24–27
Midgley DC (1963) Application of modern planning concepts to the development of the Orange River. S Afr J Sci 10/1963 59(10):461–467 (Orange River Development Conference, Bloemfontein, 25/26 April 1963)
Midgley DC (1965) Water and power for South Africa's expanding economy: a glimpse into the future. Witwatersrand University, Johannesburg, 1965, 19 p (reprinted from "EASA" March 1965)
Penn N (1995) The Orange River Frontier Zone, c. 1700–1805. In: Smith AB (ed), Einiqualand: Studies of the Orange River Frontier. UCT Press, Rondebosch, 1995, pp 21–96
Republic of South Africa (1962) Report on the proposed Orange River Development Project (ORDP). WP. X-'62

Republic of South Africa (1965) Report on the Kakamas Government Water Work and the Proposed Betterment of the Canal System. WP. W-'65

Republic of South Africa (1966) Report on the Upington Islands Government Water Scheme. WP. H-'66

Republic of South Africa (1967) Report on the proposed extension of the Buchuberg-Karos government water scheme. WP Y–'67

Republic of South Africa (1997) White paper on a National Water Policy for South Africa, WP 04-'97. Government Printers, Pretoria

Republic of South Africa (1998) National water act, Act no 36 of 1998, Government Gazette, 26 Aug 1998, Government Printers, Pretoria

Robbins D (1986) The 29th parallel: a South African journey. Shuter & Shooter, Pietermaritzburg, 1986, 226 p

Ross R (1975) The Kora wars on the Orange River, 1830–1880. J Afr Hist 16(4):561–576

Swanevelder CJ (1981) Utilising South Africa's largest river: the physiographic background to the Orange River scheme. Geojournal 2(Supplementary Issue):29–40

Turton AR, Meissner R, Mampane PM, Seremo O (2005) A hydropolitical history of South Africa's international river basins, Pretoria, WRC Report No. 1220/1/04

Union of South Africa (1917) Report of the director of irrigation for the period 1st April, 1915 to 31st March, 1916. Government Printer, Le Cap, 1945 (U.G. 29-'17)

Union of South Africa (1930) Report of the director of irrigation for the period 1st April, 1928 to 31st March, 1929 (U.G. 9-'30)

Union of South Africa (1945) Report of the commission of enquiry Kakamas Labour Colony. Government Printer, Le Cap, 1945 (U.G. 14-'45)

Union of South Africa (1937) Report to the director of irrigation, 1/04/1936-31/03/37. Government Printer, Pretoria, 1937

Wilcox AR (1985) Great river: the story of the Orange River. Drakensberg Publications, Winterton (Natal), 1986, 112 p

Worster D (1985) Rivers of empire: water, aridity, and the growth of the American West. Oxford University Press, Oxford, 1992, 402 p

Part III
Global to Local control? Glocal Actors

The Baghlan Oasis in Transition—From Autocratic Modernisation to Contested Spaces

Hermann Kreutzmann and Stefan Schütte

Abstract The Baghlan oasis is taken as a case in point to analyse the drying-up of a swampy, malaria-infested river basin in order to convert it into a "modern" agricultural production zone. The process was initiated by the Afghan kings in the first half of the twentieth century with the support of German planners and engineers, who engaged in establishing cotton and sugar production areas and the industrial processing of their crops. The implementation of modernist concepts in agricultural production made the oases of northern Afghanistan symbols of a future Afghanistan. This paper analyses the developments in the aftermath and focuses on the renaissance of similar concepts after 9/11 and the downfall of the Taliban. The transition from monarchy to republican rule, followed by subsequent turmoil and civil unrest, has affected the functioning of production in the river oases.

Keywords Afghanistan · Cotton · Sugar · Amelioration · Irrigation · Industrialisation

1 Introduction

Key concepts in agro-based developments adopt irrigation as a strategic means for productivity growth and modernisation. Northern Afghanistan experienced these developments during the first half of the twentieth century, when the Kabul-centred monarchy tried to broaden its revenue base in the north of the country. River valley oases were created and expanded north of the Salang Pass in the Qunduz River

H. Kreutzmann (✉)
Human Geography, Department of Earth Sciences,
Centre for Development Studies (ZELF), Institute of Geographic Sciences,
Freie Universität Berlin, Malteserstrasse 74-100, 12249 Berlin, Germany
e-mail: h.kreutzmann@fu-berlin.de

S. Schütte
Centre for Development Studies (ZELF), Freie Universität Berlin, Berlin, Germany

E. Lavie and A. Marshall (eds.), *Oases and Globalisation*,
Springer Geography, DOI 10.1007/978-3-319-50749-1_7

Basin (Figs. 1 and 2). The trans-mountain road to the north was the symbol of establishing a permanent, all-weather link between the centre of power in Kabul and the centre of modern agriculture in the Qunduz River Basin. Beginning in 1919, the infrastructure plans of Amir Amanullah, who initiated several reforms during his ten-year reign, were the prerequisite for subsequent development programmes and contested spaces that were carried out by his successors.[1] In this paper, we explore the nexus of early oasis and infrastructure establishment, and the competition and conflict, following post-9/11 developments, of re-building a sound infrastructure and promoting a viable economy. The joint ventures of the present agribusiness partners are linked with the tradition of Afghan–German cooperation in the founding phase. Import substitution in the sugar sector was the aim and vision of promoting this joint venture. Emphasis is placed on the actors and the context of introducing "modern" agriculture into a "hostile" oasis environment. The nature of the challenges has changed over time while the modernisation paradigm seems to have survived any vagaries in political rule and bilateral cooperation.

2 Environmental Conditions in the Qunduz River Basin

Afghan rule by the Durrani dynasty (1747–1973) was first established in the Pashtun heartland with Kandahar, Kabul and Herat as its centres of power. Afghan Turkestan (Fig. 1), north of the Salang Pass (3878 m), represented a contested space where local potentates tried to follow a separate and independent course in the area between the Afghan and Turk emirates. During the second half of the nineteenth century, Amir Sher Ali Khan and Amir Abdur Rahman managed to incorporate this part of the Bactrian plains as an integral part into their dominion. In the aftermath, a number of Pashtun immigrants were resettled in Afghan Turkestan with the Qunduz River Basin (Fig. 2) becoming the arena for the establishment of river-supplied irrigation oases. A favourable climate was accompanied by favourable edaphic conditions with suitable soils augmented by loess deposits and colluvial sediments. The major oasis towns of Pul-e Khumri, Baghlan, Qunduz and Khanabad are located in an altitudinal range between 390 and 640 m, before the Qunduz River drains into the Amu Darya at 320 m a.s.l. Some of the low-lying oases, such as Baghlan-Ghori and Taloqan, form basin-lake structures while others are spread out along the major rivers. The Hindu Kush mountain ranges (2100–3100 m) frame the lowlands and supply the irrigation water that is provided through the drainage systems of the Khanabad and Qunduz rivers.[2] In contrast to a comparatively high rainfall at the Salang Pass (approx. 1200 mm), annual precipitation

[1]Cf. Gregorian (1969), Jäkel (1977), Poullada (1973: 141).

[2]The Khanabad (Farkhar, Taloqan) and Qunduz (Baghlan, Pul-e Khumri, Surkhab) rivers are known under a variety of names in different sections; cf. Grötzbach (1990: 266–267), Michel (1959: 76).

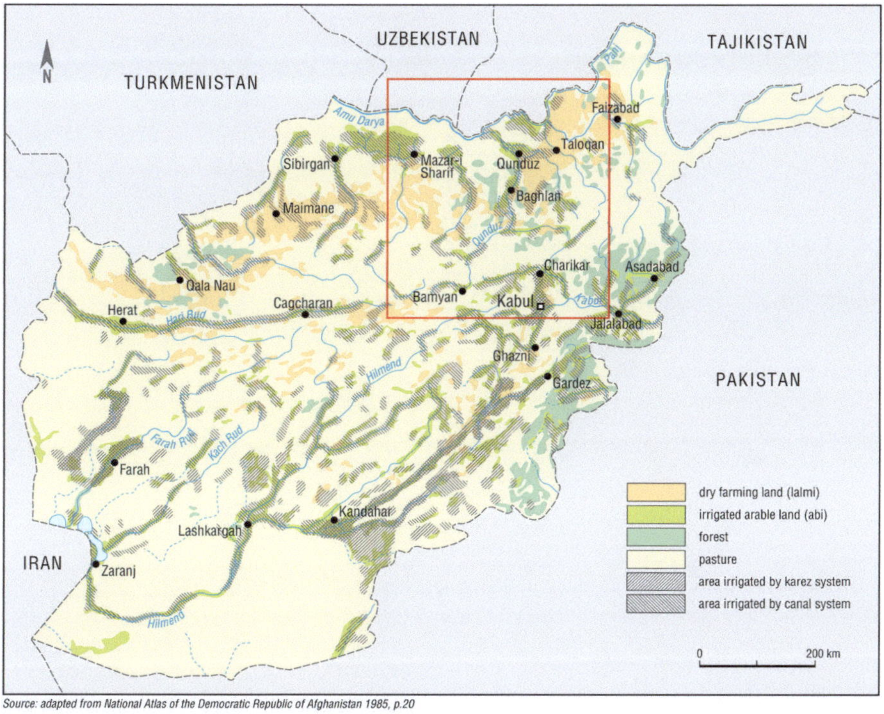

Fig. 1 Irrigation in Afghanistan. Canal irrigation in river oases is the dominant feature in Afghanistan while karez irrigation is mainly found south of the Hindukush. Design: Hermann Kreutzmann, cartography: Bernd Hilberer

is generally low with values varying between 240 and 390 mm in Baghlan (550 m) and Qunduz (455 m), peaking in late spring and early summer.[3] The area represents a region where rain-fed agriculture reaches its threshold and where a high variability in precipitation affects the likelihood of stable harvests. The summer temperatures of Baghlan and Qunduz are among the highest of Afghan climatic stations; in such a setting, augmenting irrigation can make a major difference to crop cultivation. By 1966, the "Kunduz Khanabad Irrigation Study" estimated the overall irrigated area in the Qunduz River Basin as 189,830 ha, which made up more than half of the cultivated land.[4] Efforts were made to establish a "modernised" agriculture, aimed at introducing technological means of mechanisation and agro-industrial and chemical inputs to enhance the volume output in a region where about five per cent of Afghanistan's population lived. The focus was placed

[3]For the environmental properties of northern Afghanistan, see Grötzbach (1972: 21–51, 1990: 266–286); Humlum (1959: 154–156), SOGREAH (1966, I: 23).

[4]SOGREAH (1966, I: 6, 24).

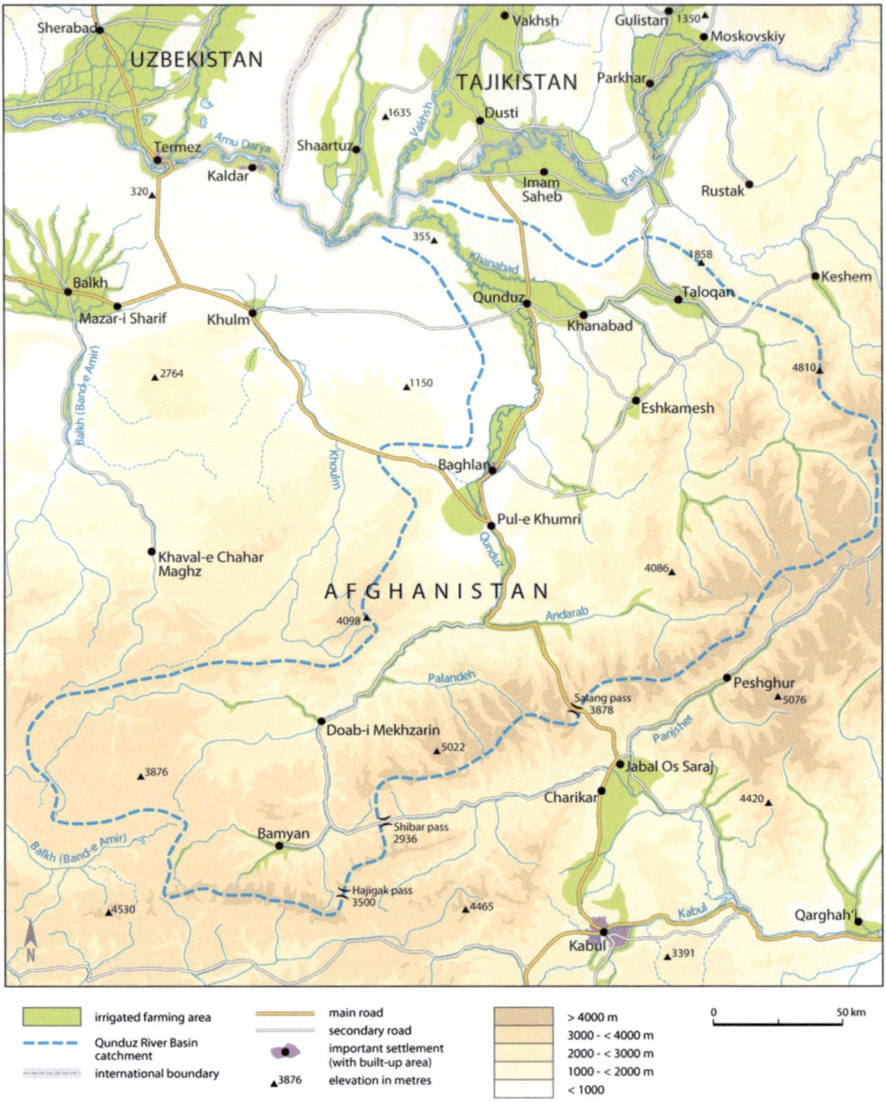

Fig. 2 Qunduz River Basin contains the canal irrigated, modern river oases that were conceived as the centres for modern agriculture in the cotton and sugar sector. The catchment was converted from swampy low-lying areas into canal colonies along the rivers. Design: Hermann Kreutzmann, cartography: Bernd Hilberer

on the highly productive cultivation of sugar beet (*Beta vulgaris*) and cotton seeds. Both commodities contributed less to the harvested yields than the "traditional" crops of wheat, barley and rice, but were regarded as strategic crops that covered about one-third of Afghanistan's cotton production, and all of its sugar beet cultivation to date, which equalled one-third of Afghanistan's sugar imports.[5] More than two-thirds of Afghanistan's cotton processing took place in the region. The founding of "modern" oases followed a strategy that entailed significant interventions in a previously malaria-infested swampy area. Barfield (1981: 29) described the major transition that was initiated in the twentieth century: "Until the mid-1930s the ecology of the lowland valleys remained exactly as described by British reports in the 1830s—it was an area of malarial swamps. But in the last fifty years it has become one of the most productive agricultural areas in Afghanistan". Major constructions for drainage and waterworks had to be established prior to the resettlement of mainly Pashtun nomads and Turkish-speaking refugees, who were compelled to adhere to certain cultivation schemes and who were confronted with a new crop from temperate climates. The introduction of sugar beet can be regarded as the symbol of a development effort that was embedded in the Afghan–German collaboration. After the initial military cooperation and efforts in modern town planning, agro-industrial enterprises played a significant role.

3 Modernisation as a Joint Venture: Afghan–German Collaboration

Prior to the agro-economic and technological cooperation between the Afghan monarchy and Germany in the Qunduz River Basin, there had been a military–industrial cooperation that had begun in 1903 when the first German-built weapons were delivered to Afghanistan, after the British director of the Kabul arms factory was replaced by a German. Gottlieb Fleischer was a former employee of Krupp and became the tragic victim of a devious murder when he left the country after fulfilling his duty.[6] The German intervention in the "Great Game" was without consequence up until the Asia Convention of 1907, but in its aftermath, and especially when Afghanistan gained a sovereign stand as a result of the third Anglo-Afghan War of 1919, various activities stimulated Afghan–German collaboration in order to attempt a more independent diplomatic and political performance by the Afghan monarchy.[7] In 1921, Amir Amanullah sent a delegation to Germany to negotiate with German President Ebert the possibility of signing up engineers and technicians for the development of the Afghan primary and secondary sectors. Consequently, engineers and craftsmen started building the new capital Dar-ul Aman. A year later, the reformist Amir Amanullah sent his Minister of War, the later King Nadir Shah,

[5]SOGREAH (1966, I: 6, 23).

[6]Grobba (1967: 11).

[7]Kreutzmann (2013a, b, 2014).

on an inspection tour of Qataghan and Badakhshan to explore the economic potential and the related socio-economic conditions, among other objects of interest.[8]

However, it took the ousting of Amir Amanullah and the murder of Amir Nadir Khan before a systematic development began that supported a joint effort of German expertise paired with investment by Afghan entrepreneurs in *spinzar* (white gold = cotton) and sugar beet. The autocratic ruler Nadir Shah supported the Governor Shir Khan in draining the amphibious lands by applying forced labour and in providing irrigated land for settlers.[9] His successor on the throne, Amir Zahir Shah, continued the amelioration programme in the north and formalised the cooperation with European experts. In 1935, a joint development programme was negotiated between both parties in Kabul and Berlin. The agreement was finalised in August 1939 in Berlin and included a German commitment to infrastructure development, financing agricultural and industrial investments of the order of 55 million Reichsmark, which were supposed to be compensated by Afghan export commodities.[10] A number of projects were implemented and mediated by the Afghan National Bank (Bank-e Melli): Khanabad was developed as a major rice bowl, Qunduz for cotton, which was introduced to the area in the early 1930s, and Pul-e Khumri[11] for a mix of cotton and sugar beet, while Baghlan became the centre of sugar beet cultivation and processing a little later (see Fig. 2).[12]

4 Baghlan Fobrica—Centre of Agro-Industrial Sugar Production

Import substitution was the driving force for the establishment of sugar beet cultivation in the Baghlan oasis (Fig. 3). Then and now, it was stimulated by the belief that a growth centre would create spin-off and trickle-down effects for a general process of regional development, which would interlink infrastructure expansion, energy generation, irrigated agriculture and industrial processing and production. In addition, employment for the recently settled and growing population groups and the transition to a market-oriented economy were envisaged. The founding of the *Deutsch-Orientalische Handelsgesellschaft* in 1923 can be interpreted as the beginning of a new phase "for the purpose of promoting the import and export trades with the lands of the orient, principally with Afghanistan".[13] Implementing

[8]Koshkaki (1979).

[9]Cf. Ali (1946: 25–27), Barfield (1981: 29–30), Pikulin (1956: 230).

[10]Grobba (1967: 57–58).

[11]The Pul-e Khumri Textile Company's cotton weaving and spinning mill was established between 1938 and 1945 with the help of Siemens technology (Michel 1959: 84–85).

[12]SOGREAH (1966, II: VI–10).

[13]Nicosia (1997: 246).

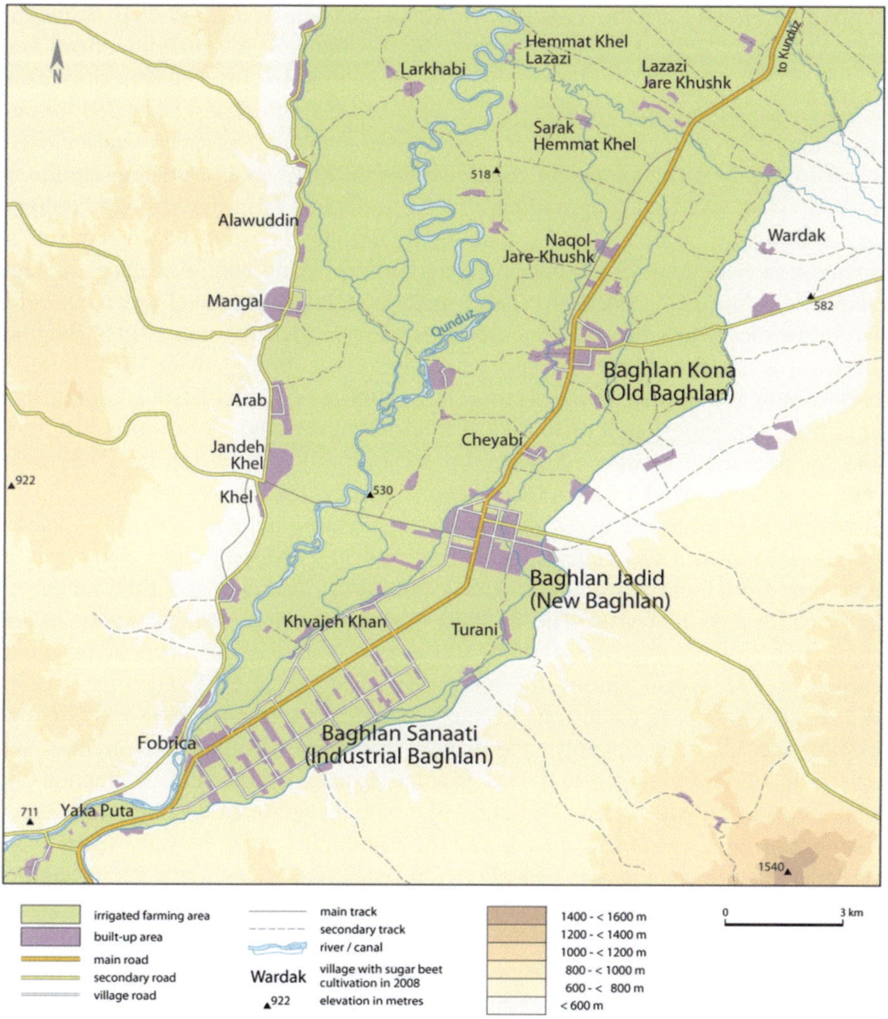

Fig. 3 Baghlan Fobrica, the immediate oasis location for the supply of sugar beet roots to the Baghlan sugar factory. Design adapted and modified from Kreutzmann and Schütte (2010: 5), cartography: Bernd Hilberer

the 1935 bilateral agreement, the Baghlan region was identified as an experimental ground for new agro-technical innovations. The first impetus for industrial sugar production in Afghanistan originated from the formation of the Baghlan Sugar Factory through the Afghan National Bank (Bank-i-Melli) in 1936/37. Louis Dupree commented: "The Bank-i-Melli served as a centre for capital accumulation, and investments flowed back from the bank to northern Afghanistan and contributed greatly to the creation of almost all pre-World War II industrial development at Pul-i-Khumri, Kunduz, and Kabul".[14] A catalytic effect was attributed to the bank's investment strategy, promoting the involvement of Afghan entrepreneurs in industrial activities that were based on the latest imported foreign technology.[15] At the same time, a socio-demographic experiment was closely linked to the oasis development. Close to the factory, about 600 households of Pashtun, Tajik and Uzbek agriculturists were settled to form the irrigation colony; big landholdings were rejected as the small farming households were meant to provide a new element in a modernised agriculture. In no way were they free and independent in making decisions about their production. By royal order, all colonists were meant to cultivate the crops that were needed for processing. Baghlan Fobrica was no exception to the rule with its emphasis on sugar beet, while in other oases cotton was the main cash crop. The initial plans calculated a compulsory sugar beet cultivation area of one-third of the Baghlan and Ghori oasis but, over time, the pressure exerted resulted in only one-fifth of the harvested area.[16] Farmers were even encouraged to grow beet beyond the mandatory 20 per cent limit by providing them with incentives such as credit facilities and attractive subsidies for agricultural inputs. Crop cultivation commenced in 1940 with about 8400 tons of sugar beet from approximately 1100 ha (5800 *jerib*), which enabled the factory to run for 23.5 days that year. The capacity was estimated as using 60,000–70,000 tons during a processing period of 90–100 days.[17] The maximum beet acreage was reached in 1975 when, during the longest campaign ever (175 days), an area of 24,131 *jerib* (five *jerib* equals one hectare) of dedicated fields yielded 108,046 tons of sugar beet.[18] Consequently, the factory has rarely fulfilled the dreams and aspirations of its planners; it took two decades to reach full capacity production and the peak period was between 1962 and 1979 with a significant decline in the 1980s when campaigns lasted less than a month. Although it was the only sugar plant specialising in beet processing in all Afghanistan, its productivity varied significantly. Afghanistan never reached the goal of substituting half of its sugar imports; rather one-sixth was achieved during the prosperous 1970s. Nevertheless, two-thirds were contributed by sugar beet and almost solely from Baghlan and the Ghori Plain (north-west of Pul-e Khumri and adjacent to Baghlan). Already in the mid-1950s, Michail Pikulin confirmed that the textile factory in Pul-e Khumri and the sugar factory in Baghlan had developed into the major industrial plants of Afghanistan.[19]

---

[14]Dupree (1973: 472).

[15]Grötzbach (1972: 69), Rhein and Ghaussy (1966: 65–72).

[16]Grötzbach (1972: 69, 155), Michel (1959).

[17]Michel (1959: 100–101).

[18]The data were recorded from the table displayed in the Baghlan sugar factory during fieldwork in 2008.

[19]Pikulin (1956: 229–230).

All machinery and expertise were imported from Europe; Austrian and German engineers directed the factory, the latest equipment came from the German-occupied Czech Škoda factory, a leading manufacturer in sugar technology, fuse boxes from Zurich and cutting machinery from Stuttgart. UK firms supplied Scottish special processing equipment from Paisley and agricultural machinery from Ipswich. The aim was to substitute half of Afghanistan's sugar imports in the long run. In order to reach this goal, the autocratic rule of the Afghan monarchy cooperated with German expertise. The state was meant to force the farmers to cultivate beet by applying German agricultural extension that introduced mechanisation, mineral fertilisers, high-yielding varieties, pesticides and fungicides. Mohammed Ali summarised the expectations and the vision: "The cultivated land of Baghlan was divided in plots of 12 *jerib* each. Each farmer was given one of these plots with a sum of 3300 Afghanis as loan to purchase the necessary implements. As a result of this encouragement, cultivators from all parts of the country flocked there, and those very lands which were arid and the home of scorpions, snakes, mosquitoes and poisonous spiders, were turned into smiling fields, and have now become the centre of hopes of the whole nation, and supplies, besides cereals and cotton, all the beet requirements of the Sugar Factory. Every one now wishes to have a piece of land in this area because of its fertility and good climate".[20]

The modernising spirit found its structural expression in the layout of the state-of-the-art architecture of the factory, its premises and its additional buildings for housing the factory administrators and workers, for marketing agricultural goods called "Agropunkt" and neighbouring bazaar facilities for the modern community in "Fobrica" as the central institution of "Industrial Baghlan".[21] The modernist layout, with separate functions devoted to different areas, was supposed to be the nucleus of further development in a growing and highly productive oasis settlement. A planned settlement with housing and cultivated land was established along a rectangular street grid east of the factory (Fig. 3). Twelve *jerib* of fertile-irrigated land was distributed among farming households in the vicinity of the factory premises. This policy encouraged many farmers to inhabit the new town of "Baghlan Sanaati" (Industrial Baghlan).

The factory operated almost continuously for half a century up to 1990. Although there was a significant decline in production after the Soviet occupation of Afghanistan in the 1990s, the factory was still up and running, and closed only at the beginning of the civil war in the early 1990s (cf. Fig. 4). During the Soviet occupation, heavy fighting occurred around the factory. As a government enterprise, it was protected by Soviet tanks, but came under attack from Mujaheddin forces. Factory workers were caught between the lines, being obliged to resume

---

[20]Ali (1946: 27–28).

[21]A redrawn sketch of the original plan from 1942 is reproduced in Kreutzmann and Schütte (2010: 6) as well as a map of the state of the New Baghlan sugar factory and its surroundings in 2008.

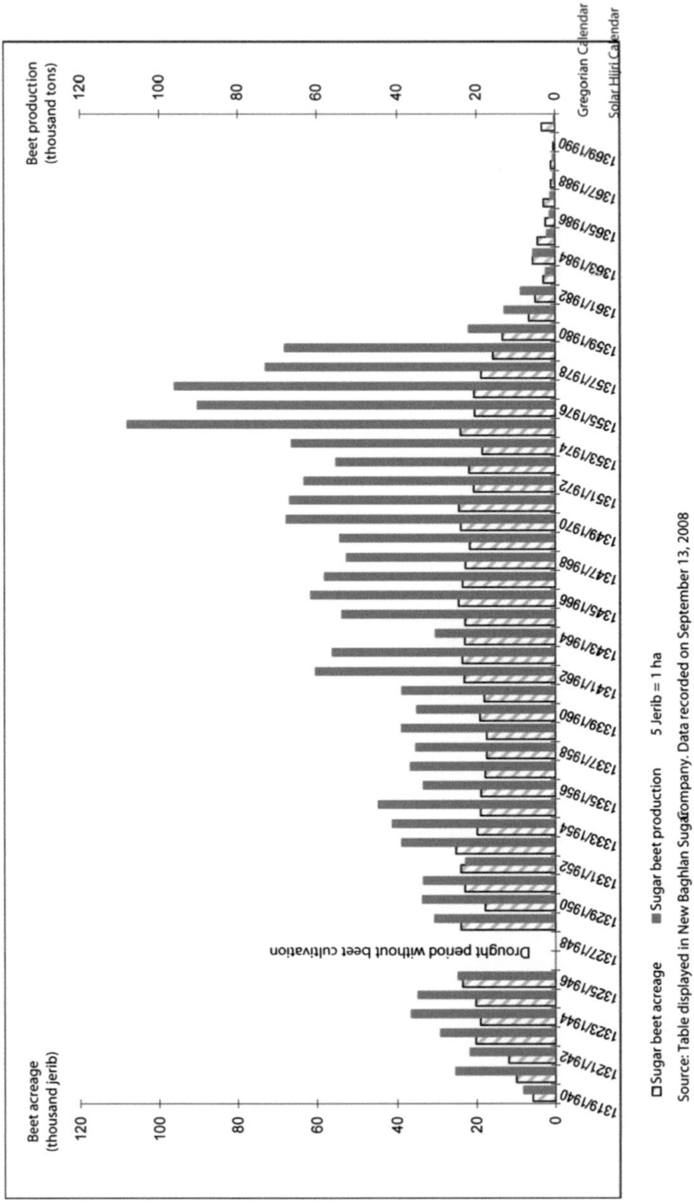

Fig. 4 Sugar beet production and acreage under sugar beet from 1940 to 1990. *Source* Reproduced from Kreutzmann and Schütte (2010)

their duties in the formally operating factory while, at the same time, having their lives threatened by anti-government forces. Workers recalled night visits, harassment and violent agitation, and many colleagues being killed by the Mujaheddin simply because they went to work in the factory. In this way, the tactics of today's insurgency in Afghanistan strikingly resemble those of the Mujaheddin. However, the sugar factory itself remained untouched, even in the subsequent civil war, and offered the opportunity of re-opening despite heavy fighting in the Baghlan oasis and the changing control of Baghlan at the crossroads that links northern and central Afghanistan. In the vicinity, the pressure on settlement space and agricultural resources has increased significantly.

5 New Baghlan Sugar Company—An Oasis in Transition?

Against this historical background, the establishment of the New Baghlan Sugar Company (NBSC) in 2004 sought to revive sugar production in Afghanistan by reconstructing the deserted factory, which had survived the civil war and Taliban rule without its equipment being completely looted (Fig. 5). This endeavour formed part of the larger project of agricultural rehabilitation seen as the backbone of the

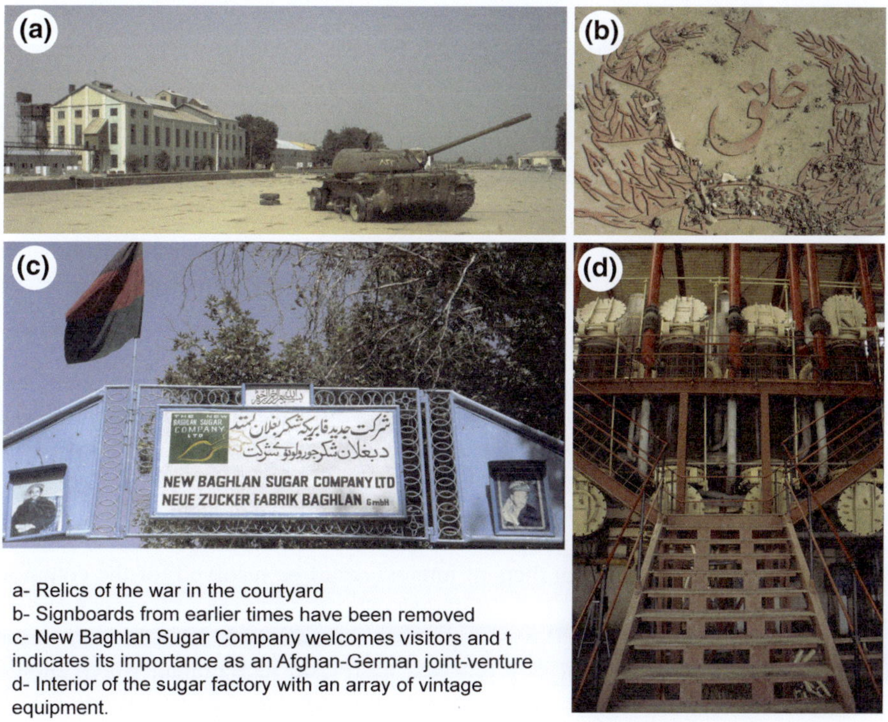

a- Relics of the war in the courtyard
b- Signboards from earlier times have been removed
c- New Baghlan Sugar Company welcomes visitors and t indicates its importance as an Afghan-German joint-venture
d- Interior of the sugar factory with an array of vintage equipment.

Source: Photographs taken by Hermann Kreutzmann in September 2006 and September 2008

Fig. 5 Baghlan sugar factory in transition

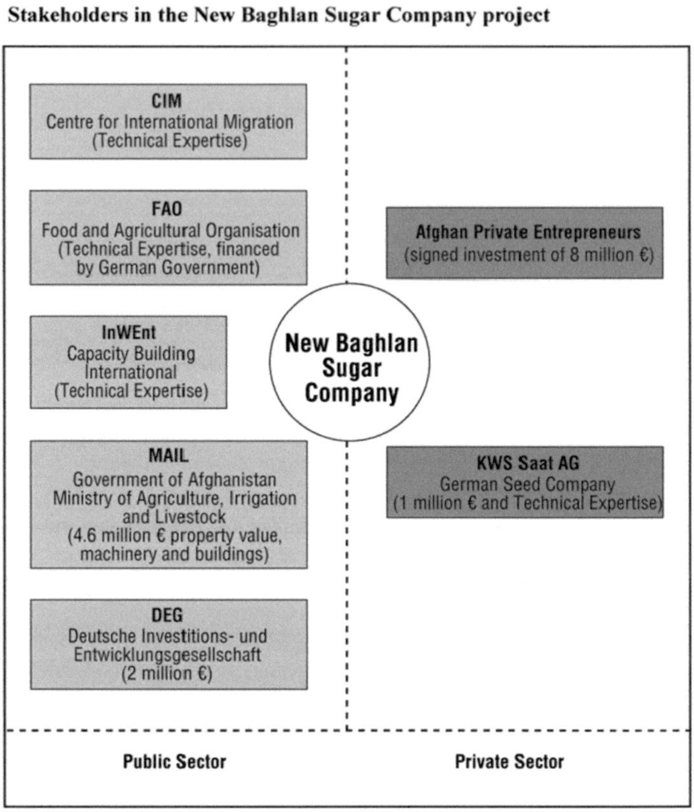

Fig. 6 Composition of the initial public–private partnership that financed and organised the re-opening of the sugar factory in 2004. This partnership has now been dismantled and further support for the venture has been phased out. *Source* Reproduced from Kreutzmann and Schütte (2010)

"alternative livelihoods" approach, the major intervention that directed international efforts to combat the war and illicit economies in Afghanistan. The basic premise of the New Baghlan Sugar Company was a public–private partnership (PPP) involving the Afghan and German governments as well as private investors from both countries (cf. Fig. 6). The contractual arrangements between state institutions and the private sector to revive sugar production were based on the promise to provide "win-win" situations where conflicts of interest could be resolved for the common good.

The explicit goal of the PPP was to provide alternatives to local farmers in the form of sugar beet cultivation as a renewed cash crop in Baghlan, which should help generate a stable income source for rural households. The transition of agricultural practices from subsistence orientation to market production was envisaged through the promotion of larger-scale commercial agriculture. The process of reconstruction was guided by the currently dominant development narratives of privatisation of state-owned enterprises and the creation of open agricultural markets serving as "enabling environments" for the promotion of rural livelihoods and the transition of the oasis into a modernised agricultural landscape. However, these "narratives of rehabilitation" (Christoplos 2007) have, in the case of the New Baghlan Sugar Company, been confronted with many internal and external problems, which will probably result in the complete failure of the reconstruction project and the abandonment of the factory by its international donors. In order to comprehend how the rehabilitation process unfolded, and finally failed, in the attempt to re-industrialise Baghlan Fobrica, a focus on the networks of actors working in the context of a dominant development paradigm is required.

Specifically, the establishment of the New Baghlan Sugar Company was perceived as contributing to combating rural poverty through the creation of new markets for small farmers and the provision of incentives for enhanced productivity, as well as consultation through agricultural extension workers. Public support came from the Afghan Ministry of Agriculture, Irrigation and Livestock (MAIL) while the German side supported the project through training programmes and professional advice facilitated by Capacity Building International (InWEnt—now part of the GIZ, the German Federal Enterprise for International Cooperation) as well as through two million Euros of financial contributions channelled through "Deutsche Investitions- und Entwicklungsgesellschaft", which kept the factory running up until December 2014. In addition, the Food and Agriculture Organisation (FAO) of the United Nations contributed technical expertise financed by German attributions. The private sector was represented, on the one hand, by four Afghan entrepreneurs from an experienced trading family, who invested substantial capital in the enterprise, and on the other hand, by the German Kleinwanzlebener Saatgut (KWS), a leading company in improved seed production seeking to re-establish traditional markets for their products in Afghanistan.

This specific group of actors had to handle high expectations while, right from the beginning, all partners struggled with many difficulties. Initially, the idea of a re-opened sugar factory tuned to modern development approaches sounded convincing to all those involved. The demand for sugar in Afghanistan is high and the supply is characterised by meagre domestic production and a high proportion of imported sugar, making substitution of imports appear a rational approach. Starting its operation in 2006, the full use of the production capacities in the renewed factory was already projected for 2009. Two thousand farmers were supposed to be engaged in the project and a significant acreage of 10,000 ha under sugar beet was targeted, predicting the creation of 11,000 direct and indirect new jobs in the Baghlan oasis. These expected prospects for a high-profit venture, as identified by a feasibility study carried out in 2003, encouraged Afghan businessmen to commit to high investments in the first place. However, these expectations were not even remotely met, and from very early on led to disillusionment. Farmers were very reluctant to commit to sugar beet cultivation and, since the re-opening in 2006, the most successful production year was in 2013 when 260 farming households produced 5155 tons of sugar beet on 951 ha of farmland.[22]

As such, the idea of a PPP as a role model for development collapsed quickly and the Afghan businessmen withdrew their funds early on, after realising that their trust in the early assessments of German experts promising a golden future for the venture was totally misplaced. Reasons for the ineffective functioning of the New Baghlan Sugar Company can be specifically attributed to a number of unaddressed bottlenecks and constraints. These include the aversion of farmers to dedicating their irrigated land to growing sugar beet for numerous reasons, a lack of capacity enabling the operation of the sugar factory without the support of foreign experts, and the inability of the company to acquire sufficient landholdings of their own.

More generally, the absolute commitment to the rule of markets and the downright profit-oriented outlook on behalf of the factory administration paired with their general ignorance about the technical processes involved in large-scale sugar production in an antique heritage factory have been critical for the present state of affairs. Accordingly, a quick demise of the project can be expected, which stands as a symbol for an Afghan past in which industrial production played a significant role in agriculture and rural employment. The nexus of rehabilitation policies wedded to ideas of neoliberal development and the actual practices on the ground regarding sugar beet cultivation and its industrial processing in the factory is outlined below to determine its impact on agricultural production in the Baghlan oasis.

6 Sugar and Rural Development in the Baghlan Oasis

Two sets of activities were crucial in the endeavour to promote sugar beet cultivation and keep the factory up and running: encouraging farmers to engage in beet production and supporting them in this enterprise through agricultural extension programmes, and reconstructing the factory and its more than 60-year-old equipment while, at the same time, creating the conditions to enable a smooth day-to-day running of the sugar plant. These fields of intervention seemed straightforward enough and consisted of rather common agro-industrial development practices. Accordingly, the question remains why the prospect of failure loomed large over the New Baghlan Sugar Company right from the beginning.

It seems obvious that, after complete abandonment for 15 years, the re-vitalisation of sugar production in Baghlan needed long-term commitment and, at least initially, heavy and committed subsidies and a strong focus on capacity building in order to evolve gradually into a productive enterprise. This also meant

[22]Personal communication with the technical management of the New Baghlan Sugar Company in December 2014.

that farmers would have to be offered incentive prices to encourage them to take on production—without their active contribution, the project was at any rate doomed to fail in the long run. However, these rather apparent framework conditions were not taken into consideration in the day-to-day running of the re-opened factory. It seems that the responsible partners saw the venture as a self-propelling undertaking with prospects for immediate success, without having taken into account the specific situations of rural Afghanistan and the rationalities of rural farmers when deciding about which crops to cultivate on their lands. The profit-oriented outlook, with the focus on the rule of markets that is congruent with current development thinking, apparently turned out to be a cul-de-sac for the enterprise.

Farmers hesitated to engage in beet cultivation mainly for economic, social and technical reasons. Without the provision of incentive prices, sugar beet cannot compete with grain crops such as wheat and maize. In addition, preparing the field for sugar beet requires comparatively high input investments in mineral fertiliser, high-yielding seed material, fungicides and pesticides, seedbed preparation, irrigation and maintenance, and a manual labour force. These inputs are partly provided by the factory, but are paid by the farmers and accounted for after harvesting. With the looming risk of meagre harvests paired with low returns for their crop, this policy risks farmers falling into debt with the factory. This is especially true as sugar beet is a high maintenance crop, requiring daily care and a consistently high work input. The crop is very fragile, demanding daily removal of weeds and very careful watering. These factors are all the more important in that only good quality beet can be delivered and used for the production of white sugar. However, many farmers have expressed dissatisfaction with the technical support they receive from the factory as their only customer and they cannot meet the required cultivation standards. There have been notable exceptions; for example, when skilled farmers engaged in labour-intensive care of beet fields were able to fend off the infestation of diseases and achieve higher productivity. Nonetheless, these rare cases show that high yields are possible, which are invariably needed in order to achieve economic benefits. The average productivity is hardly sufficient to make up for the high investments in farm inputs. In addition, farmers have also expressed discontent with the *modus operandi* of beet cultivation and the lack of support they receive. Many apparently withdrew their acreage for these reasons, complaining about a lack of consultation with local institutions and the reluctance of the factory management to engage in participatory decision-making. For instance, in 2006, a large demonstration of farmers in front of the factory gates attempted to draw attention to the conditions that would enable them to engage in beet production, but this show of public discontent did not lead to a negotiated and mutually accepted solution. Here, an opportunity was missed to engage in dialogue and find solutions to the problems as perceived by farmers. A closer look at a quite similar situation in the 1970s would have helped. Incidentally, a World Bank report from 1975 assessed great potential in Afghanistan's industrial sugar sector with high prospects for self-sufficiency, but today identifies low productivity and lack of acreage under sugar beet as major shortcomings to be urgently addressed.[23] According to the World Bank advisers, the major instrument for success would have been an appropriate pricing policy offering incentive prices that would have encouraged farmers to commit more land to beet cultivation. This, however, would have meant abandoning completely the market-based idea of running the factory, which apparently was not seen as an option by the factory administration and its donors. In fact, the productive use of available funds was monitored by the German Federal Audit Court, which repeatedly demanded proof of success of the sugar venture as measured in viable economic returns.[24] Failure to do so meant that the decision was taken not to release additional funds for the project to continue after 2014.

However, in terms of the technical running of the factory itself, things did not look so bleak. The sheer fact that the technical restoration of the factory has been successful, after 15 years of complete standstill, is remarkable in itself. This achievement was possible through the utilisation of foreign expertise and continues to be fully dependent on this external technical support, essentially provided by a single dedicated German sugar engineer, who has shared his expertise and trained employees in a variety of required skills. After seven years of constant supervision and capacity-building programmes, it seems that the complex task of operating a sugar factory that works with machines built in the 1940s has been attained under almost complete Afghan technical supervision. This is no small achievement as the sugar beet raw product has to pass through no less than 36 technically differentiated stations in the factory to be transformed and processed into refined white sugar, with maintenance and expert supervision of each of these stations being critical to the overall production cycle. Added to the difficulties of capacity building was the rather high turnover of trained personnel, who often forcefully left the factory. Thus, investments of time and effort by the technical management in capacity building are for various, mostly financial, reasons sometimes counteracted by the factory's administration. While two educated engineering personnel have been successfully trained over a long period and their services secured for the time being and a few station-experts have been effectively established, ordinary factory workers complain about working conditions. In particular, the reactivated experienced senior workers compare the present situation with the 1970s, when the conditions of employment as civil servants were perceived as very favourable. Today, however, the profit-oriented reconstruction has led to a cut-back in the most basic social security benefits. Still, the factory has managed to retain 64 permanent employees up until the last production campaign in 2014, which was further supported by 340 temporary employees. The positive effect on off-farm labour opportunities is evident, but it seems that 2014 was the last year of operation for the

[23]World Bank (1975, II: 13).

[24]Personal communication with a relevant desk worker at the German Federal Ministry for Economic Cooperation and Development, December 2014.

factory as external funds have run out and the Afghan Government, as the sole proprietor, does not seem in a position to finance the enterprise further on its own.

7 After Sugar—Rural Economies of the Baghlan Oasis

Evidently, sugar beet production has not become sustainable as a contribution to rural development and agricultural transformation in the Baghlan oasis. Although the most important personnel are still committed to the project, the outlook is bleak. What does this mean in terms of missed opportunities in shaping rural economies through agro-industrial production? The sugar factory continues to be a strong symbol of an economically viable past in the Baghlan oasis, and it was kept running, albeit on very low figures, in spite of detrimental policies shaping its operation, uninformed donors who had no prior understanding of the challenges involved in the project, and a factory administration that alienated rural farming households rather than providing incentives for beet cultivation. The business plan of the factory provided no blueprint for competition in a market economy. The failure of the project also exemplifies a failure of pure market-based development ideologies, which are seemingly "out of step" with agricultural realities in the country (Christoplos 2004). As a consequence, instead of sugar, the poppy crop appears to be on the rise[25] and, with the potential closing-down of the sugar plant, already limited rural income opportunities will become even less diversified. The failed agro-industrial transformation of the Baghlan oasis resembles another example of the development impasse in Afghanistan, which has already led to growing dissatisfaction among rural populations (Donini 2007). There might have been scope for reconciliation early on through the promotion of bottom-up approaches, which would have needed to incorporate all local stakeholders, including local farmers and factory workers, in participatory decision-making. However, this opportunity has passed, given the currently very firm donor stand, determined not to commit additional funds, and the lack of private investments. The failed modernisation of the agricultural economy in the Baghlan oasis thus can be pinpointed to a misdirected project design, which operated in the spirit of a neoliberal development policy and its sole focus on the rule of markets, but did not take into account local perceptions and priorities. It remains a mystery how project designers could have believed that the reconstruction of the sugar plant would generate profits right from the beginning. The illusive power of the dominant development narrative apparently prevented thinking about an arrangement that operated like a larger-scale development project, sustained even in the longer term by committed external support. The challenging technical reconstruction of the factory, the training of factory staff in specific skills (e.g. electrical maintenance,

[25]Personal communication with the technical management of the New Baghlan Sugar Company in December 2014.

turbine engineering, welding, lathe operating and locksmithing), the training of sugar technology experts and the long-term enlistment of farmers to cultivate sugar beet via initial price incentives in order to generate a constant supply could all have been achieved without the constraints of external financial pressures. This is especially deplorable as the funds that would have been required are rather reasonable and manageable amounts, when compared to the overall commitment of international donors to agricultural rehabilitation in Afghanistan. This ship has now sailed, and what is left after sugar has failed in Baghlan are missed opportunities at best, and a disillusioned rural population who are possibly turning to illicit crops and income opportunities, which may thwart the overall goal of rural reconstruction to improve livelihoods through the establishment of lasting opportunities in both farm and off-farm sectors.

The Qunduz River Basin in general and the Baghlan oasis in particular have become symbols of a faded impetus, which began three generations ago with great zeal and aspirations as a move to put Afghanistan on a par with their European counterparts. By draining the swamps and converting them into fertile and valuable oases, outsider interest increased and, during the subsequent conflicts, these became contested spaces that survived and were not destroyed. Surprisingly, the Baghlan sugar factory survived the battles with minor scars. The attempt to rejuvenate the historical experience of international cooperation in joint ventures in an alleged post-conflict setting had high symbolic value. For the rehabilitation of "modern" agriculture, it has been a required step in oasis development, but has remained an experiment without any implications for replication in other settings. A number of parameters have fallen short of the minimum requirements. Consequently, the oasis farmers have adapted themselves and their agricultural production to flexible strategies that support their survival.

References

Ali M (1946) Commercial Afghanistan. R. M. Bhattacharjee, Delhi

Barfield TJ (1981) The Central Asian Arabs of Afghanistan. Pastoral nomadism in transition. University of Texas Press, Austin

Christoplos I (2004) Out of Step? Agricultural Policy and Afghan Livelihoods. Afghanistan Research and Evaluation Unit, Kabul

Christoplos I (2007) Narratives of rehabilitation in Afghan agricultural interventions. In: Pain, Sutton J (eds) Reconstructing agriculture in Afghanistan. Practical action publishing, Bourton on Dunsmore, pp 165–187

Donini A (2007) Local perceptions of assistance to Afghanistan. International Peacekeeping 14 (1):158–172

Dupree L (1973/1980) Afghanistan. Princeton University Press

Gregorian V (1969) The emergence of modern Afghanistan. Politics of reform and moder-ni-zation, 1880–1946. Stanford University Press, Stanford

Grobba F (1967) Männer und Mächte im Orient. 25 Jahre diplomatischer Tätigkeit im Orient. Musterschmidt-Verlag, Göttingen

Grötzbach E (1972) Kulturgeographischer Wandel in Nordost-Afghanistan seit dem 19. Jahrhundert. Anton Hain, Meisenheim am Glan

Grötzbach E (1990) Afghanistan. Eine geographische Landeskunde. Wissenschaftliche Buchgesellschaft, Darmstadt

Humlum J (1959) La Géographie de l'Afghanistan. Gyldendal, Kopenhagen

Jäkel K (1977) Reform und reaktion in Afghanistan. Notizen zu Aufstieg und Fall Amanullahs. Mardom Nameh. Hefte zur Geschichte und Gesellschaft des Mittleren Orients 3:24–57

Koshkaki MBK (1979) Qataghan et Badakhshân. Description du pays d'apres l'inspection d'un ministre afghan en 1922 (translated into French by Marguerite Reut (=Travaux de l'Institut d'etudes Iraniennes de l'université de la Sorbonne nouvelle 10). Edition du CNRS, Paris

Kreutzmann H (2013a) Boundary-making as a strategy for risk reduction in conflict-prone spaces. In: Müller-Mahn D (ed) The spatial dimension of risk. How geography shapes the emergence of riskscapes. Routledge, Milton Park, pp 154–171

Kreutzmann H (2013b) The significance of geopolitical issues for internal development and intervention in mountainous areas of Crossroads Asia. In: Crossroads Asia Working Paper Series 7:1–31

Kreutzmann H (2014) Süße Intervention – Die Zuckerfabrik in Baghlan gestern und heute. In: Daxner M (ed) Deutschland in Afghanistan. BIS-Verlag, Oldenburg. pp 25–38

Kreutzmann H, Schütte S (2010) Persistence and change of Afghan-German economic cooperation —the case of the New Baghlan Sugar Company. Erdkunde 64(1):1–16

Michel A (1959) The Kabul, Kunduz, and Helmand Valleys and the national economy of Afghanistan. A study of regional resources and the comparative advantage of development. National Academy of Sciences—National Research Council, Washington DC

Nicosia FR (1997) 'Drang Nach Osten' continued? Germany and Afghanistan during the Weimar Republic. Journal of Contemporary History 32(2):235–257

Pikulin MG (1956) Afghanistan. Ekonomicheskii Ocherk (Afghanistan. Economic layout). Tashkent. Izd. Akademii Nauk Uzbekskoi SSR

Poullada LB (1973) Reform and rebellion in Afghanistan, 1919–1929. King Amanullah's failure to modernize a tribal society. Cornell University Press, Ithaca, London

Rhein E, Ghaussy AG (1966) Die wirtschaftliche Entwicklung Afghanistans 1880–1965. Leske, Opladen

Société Grenobloise d'Etudes et d'Applications Hydraulique (SOGREAH) (1966) Kunduz Khanabad Irrigation Study. Final Report, 3 volumes. Afghan Water and Soil Survey Authority, Kabul

World Bank (1975) Afghanistan. Opportunities for agricultural and rural development sector report. World Bank, Washington DC

Mountain-Oases Faced with New Roads: Case Studies from the Andes and the Himalayas

Emilie Lavie and Monique Fort

Abstract In the dry regions of high mountains, communities often live in small oases; relying on rivers fed by snow/glacial melt waters, villagers develop gravity-fed networks, so as to irrigate their surrounding valleys and terraces. Our two case studies, Uspallata (12 km^2) in the Mendoza River Valley of the Central Argentinian Andes, and Mustang District (almost 2000 km^2) in the Kali Gandaki Valley north of the Nepalese Himalayas, illustrate the functioning of oasis systems and recent changes brought about by the development of roads, related to their position close to international boundaries.

Keywords Roads · Mountain-oases · Systems · Himalayas · Andes

1 Introduction

As seen in the previous chapters, an oasis is primarily a fertile or green area in a (semi-)arid land, made so by the presence of water. It can be found in various topographic environments: although extended and flat desert oases are those most commonly described in the literature, oases found in arid highlands and mountains cannot be neglected due to their local significance. In fact, they are the subject of many studies, dealing with different issues: access to and distribution of water for irrigation (Kreutzmann 1988), the quality of the resources (soil and water salinity, carbon dioxide) such as in the Balad Seet Oasis in the Sultanate of Oman (Wichern et al. 2004; Luedeling et al. 2004), volcanism in the Timber Mountain-Oasis in Nevada (Christiansen et al. 1977 or Snyder and Carr 1984, for the oldest ones) and,

E. Lavie · M. Fort
UMR CNRS PRODIG, University Paris-Diderot, Sorbonne-Paris-Cité, Paris, France

E. Lavie (✉) · M. Fort
Université Paris-Diderot, 5 Rue T. Mann, CC7001, 75205 Paris Cedex 13, France
e-mail: emilie.lavie@univ-paris-diderot.fr

E. Lavie and A. Marshall (eds.), *Oases and Globalisation*, Springer Geography, DOI 10.1007/978-3-319-50749-1_8

more generally, ecosystems, biodiversity and environmental management, such as in the Ethiopian highlands (Hurni et al. 2011) and the Central-Asian mountain-oases (Sun et al. 2007; Kreutzmann 1988, 2014). However, the term "mountain-oasis" does not appear in the now-famous oasis typology of Clouet (1995). This may partly be explained by the rather small size of such oasian spaces, which are not really included in globalised trade, especially of crop products. Therefore, what are mountain-oases? Do they have special features? How can such spaces be differentiated from other irrigated highlands?

Mountain-oases do have some special features. Firstly, their position in mountain regions limits the possibilities for spatial expansion, due to the scarcity of flat areas. In fact, most mountain-oases are located on inherited flat or gently sloping terraces, either in valleys that were formerly glaciated, and/or in intermountain depressions of varying origins (tectonics, fluvial erosion, etc.). Secondly, their elevation is a significant characteristic because it controls some environmental components: the presence of glaciers, the occurrence and duration of snowfall and frost, hence indirectly the availability of water for irrigation. Thirdly, their functioning is similar to that of piedmont oases in that they "*collect water directly from mountain rivers, by water intake and canals, whose slope is calculated so that the flow is brought to the field to be irrigated by gravity*"[1] (Clouet 1995). Finally, mountain-oases can be considered "mini"-piedmont oases, i.e. isolated piedmonts within an arid mountain environment, where both the confinement and the steepness of the slopes make the circulation along routes longer and more difficult than in the plain, and may explain their relative remoteness from the main commercial routes and urban centres. Yet, they are also privileged sites where water and soil resources, though limited, contrast sharply with the adjacent, dry and sterile mountain slopes.

Our two case studies, the Oasis of Uspallata in the Argentinian Andes and the oasian archipelago[2] of Mustang District in the Nepalese Himalayas, are examples of highland oases dispersed in their arid mountain environments. Yet, sited along old trade corridors, their former systems have recently changed with the development of new roads, related to the opening of an international border. These new roads could be considered the entry points of the mountain-oasis systems into globalisation. However, the two examples differ in at least two ways. Firstly, the temporalities are not the same: the changes due to the road occurred earlier in Uspallata than in the Mustang area (where the road was completed in late 2015), and therefore, their insertion into global exchanges has not yet reached the same degree

[1]All translations from Spanish or French are by the authors.

[2]Archipelago: also see Chapter "Liwa: The Mutation of an Agricultural Oasis into a Strategic Reserve Dedicated to a Secure Water Supply for Abu Dhabi" by Cariou on Liwa. Like an island archipelago, an oasian archipelago is used to describe a group, chain or cluster of oases, generally linked to each other through a similar culture, heritage or use of resources, in a common network, forming a system.

in the two locations. Secondly, the straightforward link established between Chile and Argentina across their border cannot be compared with the sensitive geopolitical situation in which Nepal is squeezed between China and the Indian subcontinent. Although both roads were planned to develop the international trade of manufactured goods, in Mustang this project may conflict with the tourism activity closely linked to environmental conservation issues. The question then arises: how has their entry into globalisation, due to a new road development, modified these oasis systems in such a mountainous environment?

Based on fieldwork observations and a bibliographical review, rather than on a comparison between two fieldwork areas, we aim to analyse the alterations in mountain-oases following the opening of new routes: two oases with different cultures, heritages and political situations, but facing similar types of changes, with different temporalities. After a descriptive summary of these two former mountain-oasian systems (Part II), this chapter will show the transformations brought about by the development of new roads in the 1980s in Uspallata and in the 2000s in Upper Mustang (Part III).

2 An Oasis and an Oasian Archipelago Included in Continental Trade Routes

Here, we present the two "mountain-oasian" systems in their historical perspective and functioning until the recent changes of the late twentieth century. In both cases, it appears that the initial location of these village-oases is not fortuitous but related to the existence of old, natural circulation routes favoured by their relative accessibility compared to the surrounding steep mountain slopes. We show how these oasis systems, whose economy is mostly based on irrigated agriculture, husbandry and (for Mustang) seasonal nomadism, have evolved over time while still remaining open to the outside world.

2.1 Uspallata, 4000 Years of History

The Oasis of Uspallata is sited in a graben[3] between the Pre-Cordillera and the Frontal Cordillera, around 2000 m asl. This depression is drained by a series of small streams with snow-rain hydrological regimes, whose main tributary, the Uspallata Stream, merges with the Mendoza River in the south (Fig. 1b). The Uspallata Stream waters were diverted to create a small mountain-oasis of 12 km^2 in the heart of this dry environment, which receives 156 mm of precipitation a year, in both snow and rain forms.

[3]Graben: geological term for a depression between faults, resulting from tectonic movements.

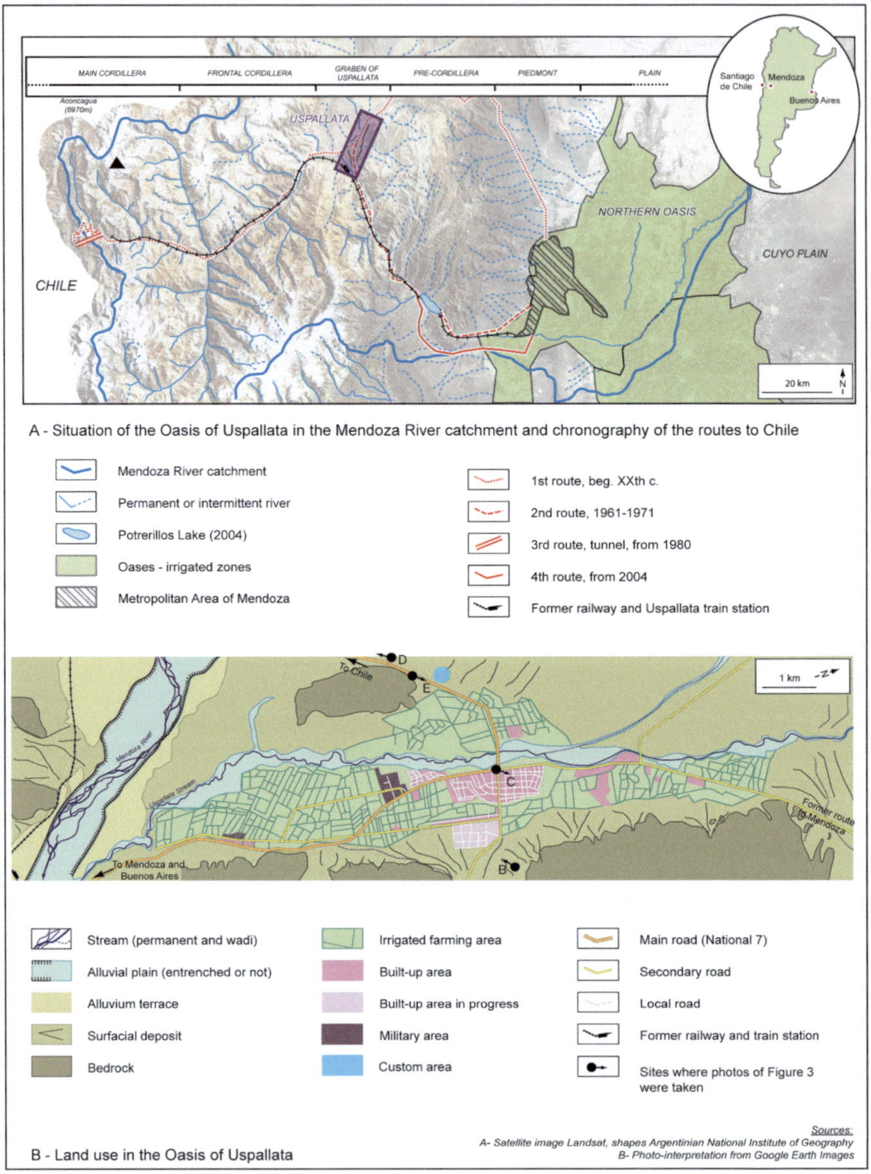

Fig. 1 Oasis of Uspallata as a node on the Andes crossing

According to some data recorded at the beginning of the Spanish colonisation at the end of the sixteenth century, a native community—named Huarpes—developed agriculture in the Cuyo region, the arid piedmont of the Central Argentinian Andes. The Province of Mendoza is now one of the three provinces of this Cuyo region.

Ponte (2006) estimates the beginning of agriculture in the Uspallata graben at around 1700 BC. By diverting small streams, people could cultivate vegetables (carrots, potatoes) and perhaps some cereals.

Then, at the end of the fifteenth century, the Incas, pursuing their Southern Conquest, arrived in Cuyo. Thus, during the 80 years preceding the arrival of Spanish colonisers in 1561, Uspallata was no longer an isolated irrigated area, but part of the road from Cuzco to the South.

During the Spanish colonisation, Argentinian mountains and piedmonts depended on the Chile Captaincy and not on the Plata Viceroyalty (which is now Argentina). Yet, at that time, the Andes and their piedmont were still the structuring axis of the roads. However, Buenos Aires could not trade directly with Europe: according to a Royal Decree, every imported product had to come from Lima. The main roads passed by Upper Peru (now Bolivia), but a more direct route from Santiago to Buenos Aires by Uspallata and Mendoza was gradually developed. Products came by boat from Lima to Valparaiso, then by road from Valparaiso to Argentina via Santiago. In the other direction, from east to west, the pampas region of Buenos Aires became a cattle-farming region from where cows were exported to Chile (Ponte 2006).

A series of rest-stops grew up in Cuyo for cattle resting and breeding. Although the main one was in Tunuyán—in the Valle de Uco, 150 km south of Uspallata (Robillard 2010)—one road passed by Uspallata. All these oases were mainly devoted to pasture crops for cattle breeding, in addition to food crops. The north–south structural axis of the Andean piedmont was replaced by a west–east axis, which is now the pan-American road from Santiago to Buenos Aires.

Argentinian National Independence in 1816, then the arrival of the railways, and with it, European migrants in Argentina (and in Mendoza in 1884: Chambouleyron 2004; Ponte 2006) did not really change the oasian landscape of Uspallata. Unlike the piedmont oases (100 km east of Uspallata following the natural axis of the Mendoza River Valley, Fig. 1a) which turned to Mediterranean crops (see Lavie et al. Chapter "The Transformation of the Oases of Mendoza (Argentina): How the Provincial Socio-Spatial Structure Was Reversed By the Crises of the 1980s and 2000s" of this volume), in Uspallata, the mountain climate and its low winter temperatures prevented a change in agriculture. Although its function as a rest-stop slowed down when Buenos Aires started to trade directly with Europe after Independence and when the railway was built, Uspallata's economic activities remained food crops and pasture. The cadastral allocation was quite similar to the one observed today, with large trees as barriers to strong winds and small plots, mainly for subsistence crops (Figs. 1 and 3d).

Yet, although the agricultural landscape did not evolve that much, the twentieth century saw a change in the role of Uspallata within the Province of Mendoza and in the relationship between Argentina and Chile.

In fact, the continuation of the railway between Buenos Aires/Mendoza to Chile had to pass by Uspallata (Fig. 1). Thus, the *transandino* operated from 1910 to 1984, even though some segments were destroyed by the flooding of the Mendoza

River. In 1978, however, it became quicker to take the road than the train, and although freight transport persisted for a few more years, it had to stop in the early 1980s after a landslide in Chile blocked the track.

Regarding the roads, the itineraries changed during the twentieth century but, as shown in Fig. 1, Uspallata remained the central traffic node, replacing the Valle de Uco, which had benefited from this role during the nineteenth Century (Robillard 2010).

Finally, its nodal situation, the flat topography of the graben in the heart of the Andes, and the existence of a stream essential for irrigation in this desert altogether made Uspallata a strategic place to control the region. It was a gathering point for the Andes Army before it crossed the Cordillera to boot the Spanish colonisers out of Chile at the beginning of the nineteenth century. Eventually, and most importantly, the National Military Force settled in Uspallata in the middle of the twentieth century, so as to control the frontier during the 50 years of alternating military dictatorship and democracy, on both sides of the Andes. As proof of its lasting presence, the Argentinian Army bought the main part of the pasture lands in 1955, playing a role in cattle breeding for personal use during the cross-border conflicts with Chile.

To sum up, the long-lasting agricultural usage was gradually completed by a nodal and military role with the Independence of the country, but this did not really transform either the system or the landscape. The new Republic in 1983 and the opening of a tunnel in 1985, both facilitating the relationship with Chile, radically changed the role and the former system of the Oasis of Uspallata.

History made Uspallata sometimes a dead end, sometimes a crossing-point. In any case, it played a strategic central role, either for trade or border control. This alternating situation depending on the political context is quite characteristic of territories close to a border, as will be seen in the next example, the Mustang oasian archipelago.

2.2 Mustang: A 3000 Year-Old Traditional Oasian Archipelago

The Mustang District (Nepal) extends over an area of 3573 km^2, with an estimated human population of 13,452 (Government of Nepal 2014), and a population density of 4.1 inhabitants/km^2 (the lowest in the country), living in oasian-type villages at elevations ranging from 2500 to 3800 m asl. The central and northern parts of the Mustang District consist of a large arid upland of tectonic origin (e.g. the Thakkhola–Mustang Graben; Fort et al. 1982). It is drained southward by the Kali Gandaki River (a major tributary of the Ganges), the very axis and waterway of the Mustang District, which makes it the natural link between south and central Asia, and the site of one of the ancient trade routes linking India to Tibet (von Fürer-Haimendorf 1975; Graafen and Seeber 1992–93).

The physiography of the Upper Mustang basin (north of Kagbeni), the former Kingdom of Lo (Peissel 1967), is quite dramatic, characterised by its asymmetry. In contrast to the eastern, barren and dissected gentle slopes, the steep, western flank rises to 6500 m and is covered by a few glaciers providing the melt waters to supply the oasis villages and irrigate their crops (Fort 2000). The Kali Gandaki River is deeply entrenched (>200–500 m) in a canyon across multicoloured sediments, which makes Upper Mustang a very attractive place despite its harsh climate. In fact, Middle and Upper Mustang are cold and windy mountainous deserts, due to their elevation (2500 m to more than 8000 m asl in central Mustang) and position on the northern rain-shadow side of the Greater Himalayas. A 6–7-month-long winter alternates with a rather short 3–4-month summer, with two very brief intermediate spring and autumn seasons. Mean annual temperatures vary between 5 and 16 °C (from north to south), but there are large disparities according to elevation, aspect and location, especially with regard to the alternating, desiccating winds. The precipitation (mainly snow) is less than 300 mm/year as measured at the Jomsom meteorological station (2700 m, Fig. 2) in Middle Mustang, but there is a

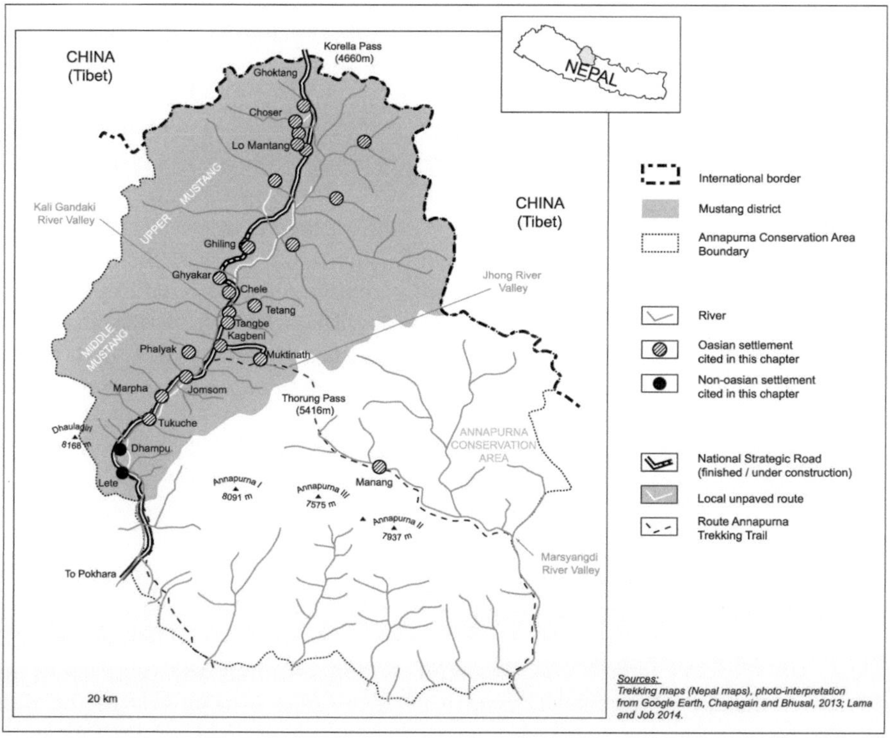

Fig. 2 Middle and Upper Mustang Archipelago, part of the Annapurna Conservation Area (ACA) Project

gradient of aridity increasing northward (<150 mm/year in Lo Mantang, 3600 m), and upward along the steep western side of the basin, limited in elevation by coldness (DHM on-line; CEPAD 2011; Fort 2014).

These different environmental characteristics explain why most of the oasis settlements are located either on the edges of wide tributary fans of the Kali Gandaki River in Middle Mustang (such as Marpha, Fig. 2), or on the gently sloping wide terrace benches built up at the end of the last Quaternary glacial period and now perched several hundred metres above the Kali Gandaki River in Upper Mustang (such as Ghyakar, Fig. 3f). This situation has made it necessary to divert waters supplied by the tributaries of the Kali Gandaki River and/or by groundwater emergences, in order to develop gravity-fed irrigation systems. Although between Tukuche and Kagbeni (Fig. 2), the village-oases sited in the arid Kali Gandaki plain are still partly overlooked by open forests of cedar or juniper trees, north of Kagbeni, at the entry of Upper Mustang, the villages appear as green islands in an otherwise wild, barren, dissected mountain landscape (such as Tetang, Fig. 3g).

The first evidence of human presence in Mustang has been provided by archaeological findings. Early occupants probably migrated from West Central Asia (Xinjiang), as attested by petroglyphs (Pohle 2000), and first settled in caves carved out of the Quaternary consolidated debris, as can be observed in many places (Simons and Schön 1998; Simons et al. 1998). In the Jhong Valley, a left-bank tributary of the Kali Gandaki River at the transition between Middle and Upper Mustang, significant remnants of early settlements (1000–800 BC) confirm at least 3000 years of permanent occupancy (Alt et al. 2003). It seems that, after clearing the natural forests (mostly pine, juniper and birch trees; Miehe 1984; Miehe et al. 2009), these early settlers developed irrigated agricultural practices based on alternating crops of buckwheat and barley in much the same way as is still done today (Knörzer 2000). Palaeo-botanical investigations also indicate the presence of other crops such as true millet and peas, and including plants derived from tropical and subtropical regions (rice, soya beans, lentils, bamboo), which indirectly demonstrates the existence of an early trading route from the southern Indian subcontinent following the Kali Gandaki Valley upstream. In addition, recent research on the spread of Buddhism from Lumbini (Lord Buddha's birthplace in southern Central Nepal) to, and connected with, the Silk Road has shown that Mustang was the most frequented of the three roads through Nepal (von der Heide 2012), because of its easy access, without high mountain passes but with a small plateau at 4660 m, the Korella Pass.

At that time and as mentioned above, the main bases of the traditional way of life in Mustang were irrigated agriculture in summer and out-migration in winter, combined with animal husbandry throughout the year, a system which, to a certain extent, still persists today. The irrigation system is composed of a series of gravity-fed canals, which usually branch out from glacial tributaries of the Kali Gandaki and run for long distances down to the villages, situated on alluvial fans or terraces (Fort 2000) (Fig. 3). The take-off canals are either carved into the bedrock (such as in Ghyakar; Fig. 3f) or more often appear as simple diversions of the

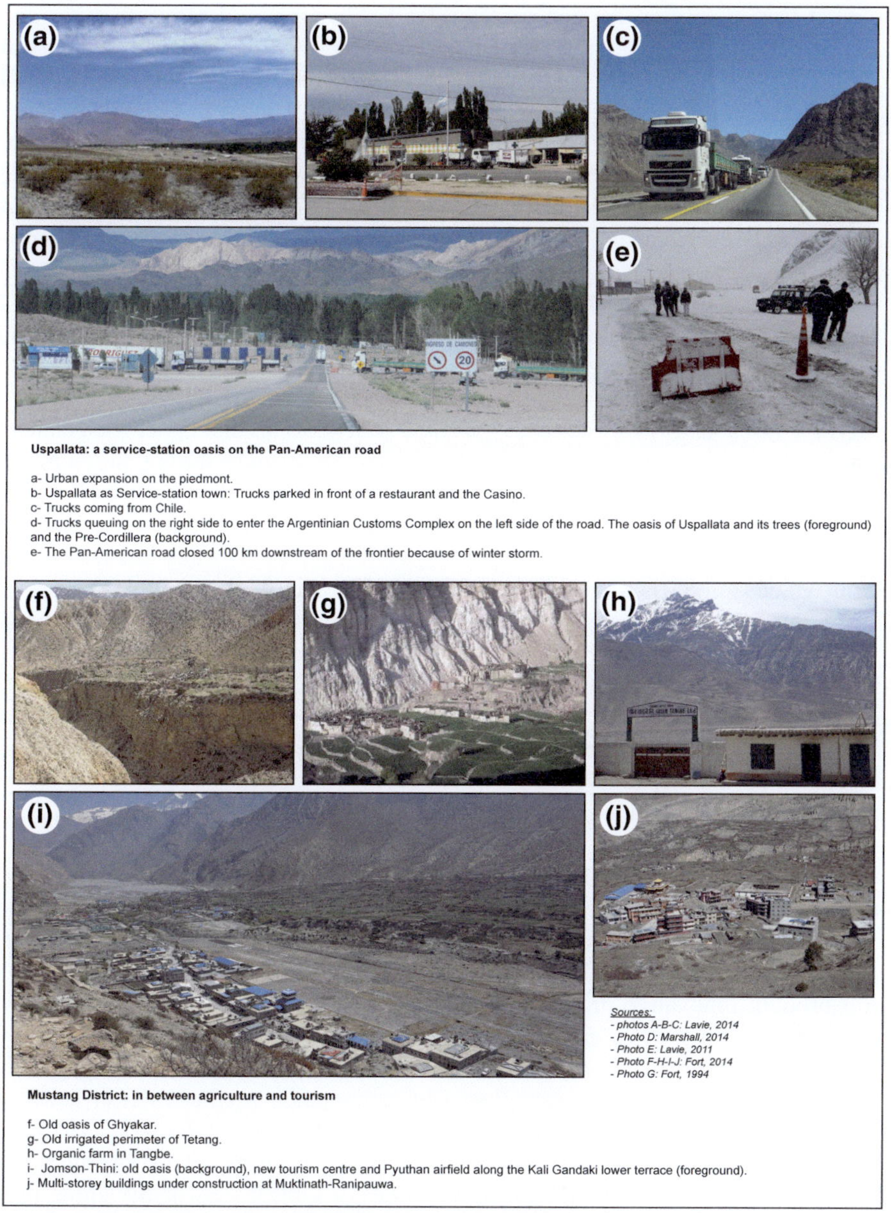

Uspallata: a service-station oasis on the Pan-American road

a- Urban expansion on the piedmont.
b- Uspallata as Service-station town: Trucks parked in front of a restaurant and the Casino.
c- Trucks coming from Chile.
d- Trucks queuing on the right side to enter the Argentinian Customs Complex on the left side of the road. The oasis of Uspallata and its trees (foreground) and the Pre-Cordillera (background).
e- The Pan-American road closed 100 km downstream of the frontier because of winter storm.

Sources:
- photos A-B-C: Lavie, 2014
- Photo D: Marshall, 2014
- Photo E: Lavie, 2011
- Photo F-H-I-J: Fort, 2014
- Photo G: Fort, 1994

Mustang District: in between agriculture and tourism

f- Old oasis of Ghyakar.
g- Old irrigated perimeter of Tetang.
h- Organic farm in Tangbe.
i- Jomson-Thini: old oasis (background), new tourism centre and Pyuthan airfield along the Kali Gandaki lower terrace (foreground).
j- Multi-storey buildings under construction at Muktinath-Ranipauwa.

Fig. 3 Transformation of Uspallata and Mustang Oases

streams, going from the streambed down to the irrigated areas. Where the tributaries originate from underground springs, they supply a limited amount of water, which needs to be stored in reservoirs, such as in Ghilling or Phalyak. The irrigation system and allocation of water over time depend mostly on socio-economic and political factors, and on the status of the land ownership and rights to water and other resources (Chalaune 2009). In Upper Mustang, the Raja used to own most of the land (and still does, even after the advent of democracy), whereas in Central Mustang the landowners belong to the highest castes. Unlike in Central Mustang where two crops a year are grown, only one crop a year is grown in Upper Mustang, primarily naked barley and buckwheat, and additional vegetables (potato, cabbage, tomato, pumpkin etc.). Agricultural work requires a high labour input and careful attention and management, including the use of natural fertilisers such as compost, manure or cattle dung.

Animal husbandry (mostly yaks, dzos, horses, mules, donkeys, sheep and goats) has always been associated with agricultural activities (Blamont 1996; Lama 2011) as it is the major source of fertiliser. It also provides many products that are used for food (meat, butter), clothes (wool, hide) and transport (yaks, horses), both on an everyday basis and with seasonal pastoral practices in the upper rangelands. Moreover, until recently, it was a major source of income during the summer transhumance across the Korella Pass with goat and sheep trading in Tibet (China), bartered with Tibetan salt (von Fürer-Haimendorf 1975). Then, after the harvest, people started to migrate south by mid-November, with Tibetan salt and wool, highly valued products sold in India. They returned to Mustang in early spring, loaded with rice and other food grains or sugar from India. It should be underlined that although winter out-migration still exists, the summer trade declined drastically after the Chinese occupation of Tibet (1959) and the closing of the northern border: the former Kingdom of Lo was declared a restricted area, Mustang lost its main source of income, and limited herding resulted in a progressive degradation of the rangelands on the Nepal side (Aryal et al. 2013).

In the following decades, several turning points occurred, initiated by decisions taken at the national level.

Firstly, in Central Mustang where a similar agricultural livelihood existed, based on irrigated crops and cattle herding to the upstream pastures of the tributaries of the Kali Gandaki, a significant change took place in the late sixties, with the setting up of the Agricultural Farm of Marpha (now Marpha Horticultural Centre). The farm brought innovative practices: for the first time in Nepal, apple, apricot and peach trees were grown, and the farm also produced vegetable seeds (Gurung 1980). This experimental farm was so successful that in the 1970s, every villager started planting orchards on their land and selling their products (fresh and dry fruits) to tourist lodges. Yet, there was a risk of over-production due to the lack of a transportation network, a concern that had already been raised in informal discussions in the early 1980s (Byers 1985). Only a limited part of the apple production could be exported, carried by porters to the closest main market of Pokhara (the second largest town in Nepal), requiring a 5-day walk.

Secondly, and in parallel with this new agricultural trend, the Annapurna Conservation Area Project (ACAP, Fig. 2) was launched in 1986 to achieve a "*sustained balance between nature conservation and socio-economic improvement*"; in particular, to reduce the pressure on open forest resources used as fuel wood and for construction (mainly lodges), and to protect the exceptional biological diversity in the rangelands of Upper Mustang. This region is a key trans-Himalayan corridor for the migration of species such as the wild ass (*Equus kiang*), the famous Tibetan Sand Grouse and the Demoiselle Crane (NTNC 2008), many of which are endangered and protected species (UNDP 2006). Managed on a participatory basis at different administrative levels, the ACAP project's aim was to protect Mustang's fragile environment while promoting economic development, a real challenge.

Thirdly, the opening of the Mustang District to tourism was certainly the main factor of socio-economic change, and a kind of compensation after the closing of the Tibetan border and the collapse of the trans-boundary traditional trade. In 1964, the former King Mahendra allowed the first foreigners to visit the Kali Gandaki Valley up to Jomsom (Sacareau 1999), then the valley was closed again from 1975 to late 1976, to avoid conflict with the Tibetan "khampas" rebels occupying the entire Mustang District. In autumn 1976, the area was again opened to foreigners up to Kagbeni and the Jhong Valley, with its famous pilgrimage site of Muktinath (for both Hindus and Buddhists coming from the entire Indian subcontinent). For the first time, the link could also be made with the adjacent, newly opened Manang Valley, hence initiating the world-famous trekking route "around the Annapurna" massif through the Marsyangdi Valley and the Thorung Pass (5416 m) (Fig. 2). This led to the building of lodges and the development of a tourism economy (food and lodging, sale of souvenirs and locally made warm clothes to visitors). At the same time, Jomsom, the District administrative headquarters with its airport strip (Fig. 3i), military camps and other governmental institutions, started growing rapidly. Thanks to tourism, the middle part of Mustang District became very successful economically, because of the proximity of the >8000 m asl Annapurna and Dhaulagiri peaks, and the overall attractive landscape with its villages of flat-roofed, white-washed adobe houses nestled among green, cereal fields and orchards, contrasting with the bare and dissected valley slopes. Eventually, this tourism boom helped to develop basic infrastructures useful for the local people (health and education services, electricity and telephone facilities; Sacareau 2009).

The fourth important step occurred in late 1992, when the "very" (Upper) Mustang was opened to foreigners, attracted both by the perspective of new trekking routes across breathtaking landscapes (the multicoloured Kali Gandaki gorges are often compared to the Grand Canyon of Colorado), and by a unique Buddhism tradition and an authentic Tibetan lifestyle (NTNC 2008; von der Heide 2012; Seiter, sd). Tourism was not immediately economically positive because visitors were obliged to be self-sufficient, i.e. to arrive with all their food, equipment, guides and porters, for trips organised by trekking agencies. Gradually, however, the possibility of selling local products, renting horses and providing seasonal employment as guides or porters began to be beneficial for the locals. In addition, NGOs started to develop solar energy successfully (Blamont and Amado 2000), a good sustainable practice to compensate for the scarcity of wood (mostly bushes) traditionally used for cooking. This new tourism-oriented evolution still did not stop winter out-migration to India with donkey or horse caravans: it only diversified the nature of the products that were imported from the subcontinent and helped to improve housing facilities (roof water tanks, solar cookers, TV, etc.).

To sum up, the Mustang District Oases archipelago consists of distinct, isolated villages where, despite a harsher climate in the upper part, livelihood practices are rather similar, and are based on irrigated agriculture and livestock, combined with seasonal migration. These oases are a network occupying a key position along the ancient trade route between India and Central Asia (Peissel 1967; Manzardo 1977), their inhabitants playing a key role along the Kali Gandaki Valley. Since 1993, the opening of Upper Mustang to tourism has progressively prepared local people for the globalisation of their local economy, and they are positively looking for a road connection linking Mustang over the Korella Pass to the existing Chinese road (UNDP 2006), a good way for them to escape the remoteness of their villages and small towns resulting from the closing of the Tibet border.

3 Bifurcation of the Systems: Development of Roads

The Upper Mustang Oases have undergone a slower evolution and more recent changes than the Oasis of Uspallata. In Mustang, the oasian system was based on summer irrigation and winter migrations with a recent opening to tourism, whereas for Uspallata, the stopover role in the Andean crossing route and the strategic military site was possible thanks to food crop cultures.

In both cases, the opening of an international route—or the improvement of an existing one—completely transformed the structure of the former system. Although the entry into globalisation was neither sudden nor similar, both areas have experienced a new position in the global trade network (for Uspallata) and, more specifically, in the international tourism network (Mustang).

3.1 Uspallata: An Oasis that Has Become a Service Station on the Pan-American Road

As shown in Fig. 1, the opening of the new route via the south of Mendoza in the 1960s (following the railway) then of the tunnel crossing the Main Cordillera in 1984 (partly using the *transandino* tunnel) and finally the new route in the early 2000s circumventing the new Potrerillos Lake that submerged the former route have progressively decreased the Mendoza–Santiago crossing time.[4] Nowadays, despite the increase in the transit, from Mendoza it only takes: 1.5 h to Uspallata,

[4]Firstly, it was very difficult to find quantitative data on Uspallata. This settlement is part of the Municipality of Las Heras (close to 9000 km² wide from the Chilean border to the heart of the Oasis of Mendoza). Most of the inhabitants live in the urban part of the Municipality [203,507 out of 207,507 in the total Municipality, according to the last 2010 Census (INDEC)], that is to say the northern part of the metropolitan area of Mendoza. Yet, data are available either at the municipality scale or at the urban area scale. Consequently, we will not give official data but figures taken from

2.5 h to the International Border and 5 h to Santiago de Chile (to which immigration formalities have to be added). Now, it takes only one day either for thousands of Argentinian tourists to go to the beach in summer, or for Chileans to go to Mendoza for some sightseeing. From Mendoza, Uspallata is also the last stop to fill up with petrol, buy souvenirs or go to a restaurant.

Moreover, although there are fewer tourists than at Mendoza or Potrerillos Lake, Uspallata offers some tourist attractions: it is the entry point to the Aconcagua Mount National Park, for trekkers coming to climb the highest mountain of all the Southern Hemisphere and the American Continent (6970 m) or for other less prestigious treks. Close to Uspallata, tourists can also visit pre-Inca and Inca petroglyphs, or geological sites with fossil stumps. Some of the routes were part of the 2009 Dakar race and now attract some four-wheel drives. Finally, the film *Seven years in Tibet* was not shot in the Himalayas but in Uspallata.

However, concomitant with the digging of the tunnel in 1984 and the new road in the early 2000s, two events changed the face of the pan-American road, as an aftermath of the return to democracy in 1983. In 1985, Argentina entered the Southern American Common Market, the MERCOSUR[5]; then, between 1988 and 1990, the organised return to democracy on the other side of the Andes, in Chile, re-opened the relationships between the two countries.

Yet, the MERCOSUR was quite a good opportunity for Brazil and Argentina to trade (each is the main trade partner of the other) and to undergo good economic growth (at least in the 1990s). Chile is not part of the MERCOSUR but has a privileged partnership, especially with Brazil. Moreover, for trucks, the fastest way to go from Santiago de Chile to the main Brazilian cities, such as Sao Paulo or Rio, is now the pan-American road via Uspallata. This small oasis remains a nodal point, not only between the two bordering countries, as presented above, but also in the whole Southern Cone. This is the situation that has really transformed the long-lasting system based on crops and military surveillance into a new one: all the economy of Uspallata is now turned towards truck traffic: a customs complex, larger-scale rest-stops (Fig. 3b), hotels, etc. According to the INDEC 2010 census, the number of inhabitants in Uspallata has risen from about 3000 in 1991 to almost 10,000 now, representing a growth rate of 17% between the 1991 and 2001 census, and 180% between 2001 and 2010. Because of the lack of land with irrigation water rights inside the oasis, new allotments are under construction in the margins of the irrigated area, that is to say without the possibility of watering gardens or trees with Uspallata Stream waters (Figs. 1 and 3a). Only drinking water will be supplied. This is not the subject of this chapter, but we can add that climatic changes and the associated decrease in snow precipitation in the watershed (as observed by remote

(Footnote 4 continued)

local newspapers (*Los Andes*, *El Uno* or *Mendoza On Line*). The majority of the given information is based on our own observations.

[5]MERCOSUR: *Mercado Común del Sur* or Southern Common Market, is a sub-regional trade bloc with customs union, comprising Argentina, Brazil, Paraguay, Uruguay and Venezuela, and associating bordering countries, such as Chile.

sensing or field surveys; Leiva 1999; Delbart et al. 2014) are challenges for the long-term water supply to this oasis, all the more so with the increase in population.

One of the problems is the excessive traffic: this pan-American road is often overloaded with cars in the summer and with trucks throughout the year. The mountain environment limits the widening of traffic lanes, so that a slow lorry reduces the speed of other vehicles (Fig. 3c) and creates a traffic jam. Moreover, the tunnel traversing the border is not wide enough to allow so much traffic. Travellers have to wait for hours on both sides to cross the Andes. Even the customs cannot sustain so much freight. The Argentinian Customs Complex, a 3 ha area some kilometres upstream of Uspallata (Fig. 1b), where trucks pass under an X-ray hangar and go through an anti-drug search, is very often full, and trucks have to wait on the other side of the road (Fig. 3d). In addition, although daily queues are quite long, they are generally worse when a winter storm, named *Viento blanco* (white wind), occurs as access to the upstream part of the watershed is closed (Fig. 3e). Private vehicles, but more often heavy trucks in transit, are blocked before Uspallata's Customs Complex. According to newspapers, on *Viento blanco* days, about 2000 trucks are waiting in the small Oasis of Uspallata (12 km^2), sometimes for a whole week. Drivers have to wait in the mountains, with negative temperatures night and day, for the re-opening of the road. Sometimes the authorities stop trucks in the plain, close to Mendoza, to prevent congestion in Uspallata.

To deal with this recurrent problem on the main route between the three most important countries of South America, already lasting a decade, the authorities and the media are discussing a new tunnel (named John-Paul II), starting from Uspallata or 20 km downstream from the current tunnel. Another project mentioned is the restoration of the *transandino* train, so as to limit road freight. The state of the remaining railways, following 30 years of landslides, makes one wonder about the feasibility of such works. One of the reasons is the funding of these road and rail infrastructures: for example, except for the tunnel fee (3 $AR = 0.35 $US = 0.30 € for cars and 15 $AR = 1.75 $US = 1.5 € for trucks), the route from Mendoza to Chile is totally free!

3.2 In Mustang

After the opening of the trekking route up to Kagbeni in 1977, then of the northern Kali Gandaki Valley by the end of 1992, the Mustang District was increasingly frequented by tourist groups, when the project of building a road from the southern Himalayas to the north took shape. The locals saw this project as a chance for development and a better livelihood, providing them with an opportunity to increase trade and cultural exchanges with the outside world. They had no perception of the potential adverse impacts of the road. After presenting the historical background of this project, we will consider the positive and negative consequences of the road for Mustang development.

After the closing of the northern border in 1959 by the Chinese Government, and after a 15-month-long trade and transit embargo imposed by the Indian Government in 1989–1990 (Ramirez 2009), the Government of Nepal suddenly became aware of the landlocked situation of the country. In order to expand the road network across the mountains and improve access to rural areas and market sectors, the Department of Roads planned in 2004 a Strategic Road Network along the main Himalayan valleys (NTNC 2008). It is in this context that the Beni-Jomsom road (linking the Kali Gandaki Valley to Pokhara) was first constructed with the help of the Nepal Army. The next step was to establish a road link between the Nepal national highway network and the road from the Tibetan Autonomous Region of China. By the word "road" we do not mean a metal/tarred road with heavy traffic, but a dusty, carriage track for trucks and all-terrain vehicles. The construction was initiated by the Mustang Development District Committee as a response to demands from farmers and traders of Mustang district, and the project was supported by local politicians and approved at the national level (Lama and Job 2014). The first part of the Kali Gandaki road was opened in late 2007 from Ghasa in the south of Mustang District up to Jomsom, but since then it has occasionally been blocked by landslides or debris flows during the summer monsoon. The Upper District part of the road was opened in late 2015, and vehicles can theoretically go up to Ghoktang, just 1 km south of Korella Pass, at the Tibet–China border, strictly controlled by the Chinese. With the full opening of this strategic road, Mustang District is no longer a remote area operating at the pace of mule caravans and trekkers, as can be observed from the rapid evolution of the economy in the last few years. Lo Manthang, the capital city of Upper Mustang, can now be reached by truck from Kagbeni in less than one day, while Jomsom can be reached within a day's drive from Kathmandu.

The impacts of the road are multifold and sometimes contradictory.

Firstly, the road is an important factor in triggering changes and, as pointed out by Lama and Job (2014), "*the road connectivity has intensified the integration of the local economy with the global*". This is particularly true of the evolution in tourism and farming, which both provide new sources of income to the local population.

More facilities are now offered to tourists, who can either trek or be transported rapidly up to Lo Manthang and the surrounding sites, attracted by the exceptional landscapes, the Buddhism cultural heritage and, more generally, by ecotourism and wilderness (Lama 2011; Devkota 2011). Pilgrims to Muktinath (Jhong Valley) can now reach the temples in less than 2 h from Kagbeni. Multistorey, concrete buildings are now part of the new landscape (Fig. 3j). Running tourist lodges is mostly beneficial to the richest landowners who can invest money to improve the accommodation with facilities that meet western standards (toilets, hot showers, electricity). Yet, it also offers more job opportunities to socially disadvantaged people, such as service activities in the lodges and guiding services for trekking groups (NTNC 2008; Poudel 2008).

The road opening from Kagbeni southwards now provides a good way to export horticultural products (mainly vegetables and fruits) to the main markets of Pokhara and Kathmandu, as well as to India. Thanks to the ideal climatic conditions (sunshine most of the year and cold winters enabling vernalisation), the "orchard" trend, initiated a few decades ago in Marpha, has spread rapidly in Upper Mustang. Fruit trees (mostly apple) have developed on any arable land available and farmers have built large water tanks accordingly, as observed in Pangling and Chele (Chapagain and Bhusal 2013), to ensure sufficient irrigation hence sufficient yields and profit (in 2013, the price of 1 kg of apples—120 NR—was 20 times higher than in the period before the opening of the road; Fort 2014). Transport times and costs have been reduced by about 75%, indirectly causing a decrease in the number of horses and other pack animals reared in Upper Mustang. New developments, such as the Organic Apple Farm south of Tangbe (Fig. 3h), should be even more profitable (it is too recent to be productive yet). More generally, the road has led to a series of changes in land use and cropping patterns, hence an increase in the demand for agricultural labour, all the more essential as a large proportion of the young, active people (25–40 years old) have emigrated either to other regions of Nepal or abroad.

However, the success of this road does not mean a real trans-boundary exchange. In fact, while the road has improved the accessibility of Upper Mustang to the external world, it has mostly increased trade with India and does not represent a real trans-Himalayan link between India and China. The explanations rest on both sides. On the Nepal side, there seems to be no policy at any level governing the development of the road (UNDP 2006), and the Government of Nepal requires trucks coming from China to pay taxes if they go beyond the Nepal checkpost and customs, whereas foreigners coming from the Mustang Nepal side are required by the Chinese authorities not to go beyond Ghoktang, less than 1 km from the Pass (as observed in 2014). In addition, on both the Nepal and China sides, there has been a border conflict with the local herders since 2004, when the Chinese authorities erected 34 km of 2.5 m high security fencing just inside their territory along the border at the head of the Pass (UNDP 2006). Closing the border has resulted in a decline in the number of grazing stock and the overgrazing and degradation of the rangeland of Upper Mustang (Paudel and Andersen 2010; Aryal et al. 2013) together with a significant loss in biodiversity (Lama 2011).

Other unexpected, adverse impacts are related to the environment resources and quality, which have to be considered in the context of climate change (Su et al. 2013) and a tourist-based economy. (i) Any new road brings noise and pollution, which are additional threats not only to wild species, but also to people, either locals or visitors. This is all the more important as Upper Mustang lies at the core of the ACA protected zone, and the very presence of this strategic road attracts more activities, more specifically at bus or vehicle stops: tea shops, lodges, stores of raw materials and chemicals, local food factories, carpet industries, etc. (ii) The road has also widened the range of tourists both in demography and origin: there are now many Indian and Chinese, besides the Western and Domestic tourists (Lama and Job 2014). Senior tourists are now increasing in number, with higher requirements for accommodation as well. This results in greater pressure on water resources

(Pandeya et al. 2015), all the more serious as there is still no system to either monitor water quality or treat it (absence of sewerage facilities, toilets without safety tanks, etc.). Collectively, this evolution is generating more solid waste and more air, ground, and water pollution. (iii) Whereas in Upper Mustang tourism is growing rapidly, in Middle Mustang, the road is now considered the main factor responsible for the reduction in trekking tourism, which was still the main source of income ten years ago. Trekkers do not like to walk along a noisy, dusty and windy road, and they are now more attracted towards Upper Mustang, and most of them no longer stop in places such as Tukuche or Marpha (Middle Mustang) where most of the lodges are now almost empty. (iv) In addition, the new road is more vulnerable to natural hazards than a foot trail and can be seriously damaged during extreme events (avalanches, landslides, debris flows and torrential floods) as already observed in Middle Mustang (Fort 2014): the losses induced by the interruption of traffic and the road maintenance might considerably increase the costs of the road for such a country as Nepal. (v) In contrast, water scarcity, together with the attraction of the road, has encouraged some local communities to move closer to it (Bernet 2012).

Ultimately, whatever the decisions already made to adjust to the new socio-economic changes generated by the strategic road, the long-term trend of climate warming and uncertainty about water resources in Mustang raise the issue of sustainable development in both farming and tourism activities, and question the preservation of this unique cultural and natural environment.

4 Conclusion: Facing a New Organisation

The two examples developed here are finally quite different in their transformations:

- In Uspallata, a self-sufficient oasis for thousands of years, alternately playing the role of a rest-stop (during the Inca and Spanish occupations and with the first European migration waves) or a dead end/military centre (second half of the twentieth century), became, with the opening of the Chile–Argentina frontier and the MERCOSUR in the 1980s, a huge service station for trucks. In the end, agriculture has been overshadowed: water is mainly used to irrigate the barrier trees and for the drinking-water supply. Other products are brought from Mendoza but not produced locally.
- In Upper Mustang, oases were summer production places, with out-migrations in winter combined with animal husbandry throughout the year. The opening of the Kali Gandaki Valley to tourism then the construction of the national Strategic Road during the last decade including the perspective of trade with Tibet have partly shifted the socio-economic purpose of such spaces. In fact, the interest in tourism and orchards has been beneficial to the richest landowners who could take the “apple turn”, while most other families had no choice but to accept

service activities (renting lodges, becoming farm employees) or to migrate to Nepalese metropolitan areas such as Kathmandu or Pokhara, or abroad.

Nevertheless, although the geographical structures of these two systems are quite different, a similar bifurcation can be observed: the international route has transformed the role of these oases, their traditional actors and their position vis-à-vis the urban decision-making centres.

These oasian areas have experienced adaptations, like many other systems faced with an anthropogenic or natural disruption. The agricultural vocation in Uspallata is now secondary, and we may ask: is it still an oasis? Finally, their mountainous location close to international borders seems to be more important than the type of oasian landscape and organisation. In contrast, the fragility of the oases and their strong dependency on water and land resources for a very dense population on a small area make them particularly vulnerable to climate changes and to the associated reduction in water discharge.

References

Alt KW, Burger J, Simons A, Schön W, Grupe G, Hummel S, Grosskopf B, Vach W, Buitrago Téllez C, Fischer CH, Möller-Wiering S, Shrestha SS, Pichler SL, Von den Driesch A (2003) Climbing into the past—first Himalayan mummies discovered in Nepal. J Archaeol Sci 30:1529–1535

Aryal A, Brunton D, Pandit R, Rai RK, Shrestha UB, Lama N, Raubenheimer D (2013) Rangelands, conflicts, and society in the Upper Mustang region, Nepal. Mt Res Dev 33(1):11–18

Bernet D (2012) Moving down or not? Report of the first field visit to Upper Mustang in early May 2012 (Part I: Synthesis). Unpublished internal report. Kam for Sud, Tathali, Nepal, 66 p

Blamont D (1996) Upper Mustang's shifting animal husbandry practices. In: Miller DJ, Craig SR (eds) Rangelands and pastoral development in the Hindu Kush—Himalayas, ICIMOD, Kathmandu, Nepal, pp 171–191

Blamont D, Amado P (2000) Using photovoltaics for agricultural processing activities in Upper Mustang (Nepal). In: Grossman G (ed) Proceedings of the ISES 1999 solar world congress, vol III, pp 495–502

Byers A (1985) Resource management in the arid Himalaya: problems and prospective solutions. In: Contribution to Nepalese studies, vol 13, no 3. CNAS Tribuvan University

CEPAD (2011) District climate and energy plan: Mustang district. CEPAD Consultants (P) Limited, Lalitpur, 131 p

Chalaune P (2009) Water, wealth, and power: a case study from Upper Mustang of Nepal. Occas Paper Sociol Anthropol 11:104–125. doi:10.3126/opsa.v11i0.3033

Chambouleyron J (2004) El aprovechamiento del agua. On-line publication: www.irrigacion.gov.ar

Chapagain PS, Bhusal JK (2013) Changing water regime and adaptation strategies in Upper Mustang Valley of Upper Kaligandaki Basin in Nepal. Sci Cold Arid Reg 5(1):0133–0139. doi:10.3724/SP.J.1226.2013.00133

Christiansen RL, Lipma PW, Carr WJ, Byers FM, Orkild PP, Sargent KA (1977) Timber Mountain–Oasis Valley caldera complex of southern Nevada. Geol Soc Am Bull 88(7):943–959

Clouet L (1995) Les oasis. Mappemonde 4(95):44–48

Delbart N, Lavie E, Goma R (2014) Influence des surfaces enneigées sur les débits des cours d'eau du piémont andin argentin. Bull Assoc Geog Fran 2014-3:338–354

Devkota D (2011) Sustainable cross-border ecotourism strategy in the Tibetan Plateau region. In: Kreutzmann H, Yong Y, Tichter J (eds) Pastoralism and rangeland management on the Tibetan Plateau in the context of climate and global change. Deutsche Gesellschaft für Internationale Zusammenarbeit (GIZ) GmbH, pp 123–141

DHM (Department of Hydrology and Meteorology) Climatic records on line: www.dhm.gov.nep/climate/

Fort M (2000) Natural conditions and hazards for irrigation in the arid Himalaya of Upper Mustang District, Nepal. In: Kreutzmann H (ed) Sharing water. Oxford, pp 239–258

Fort M (2014) Natural hazards versus climate change and their potential impacts in the dry, northern Himalayas: focus on the upper Kali Gandaki (Mustang District, Nepal). Environ Earth Sci 73:801–814. doi:10.1007/s12665-014-3087-y

Fort M, Freytet P, Colchen M (1982) Structural and sedimentological evolution of the Thakkhola-Mustang Graben (Nepal Himalaya). Zeit f Geomorph NF, Suppl-Bd 42:75–93

Government of Nepal (2014) National population and housing census 2011. VDC/Municipality. 42: Mustang. Central Bureau of Statistics, Government of Nepal, 29 p

Graafen R, Seeber C (1992–93) Important trade routes in Nepal and their importance to the settlement process. Anc Nepal, 130–133, 34–48

Gurung HB (1980) Vignettes of Nepal. Sahayogi Press, Kathmandu, 435 p

Knörzer KH (2000) 3000 years of agriculture in a valley of the High Himalayas. Veget Hist Archaeobot 9:219–222

Kreutzmann H (1988) Oases of the Karakorum: evolution of irrigation and social organization of Hunza, North Pakistan. In: Allan NJR, Knapp GW, Stadel C (eds) Human impacts on mountains. Totowa NJ, pp 243–254

Kreutzmann H (2014) The Central Asian oases—from trading towns to centres of modernization. In: Proceeding of the international colloquium oases in the globalization: ruptures and continuities, Paris, 13–14 Dec 2013. https://hal-univ-paris13.archives-ouvertes.fr/OASIS_MONDIALISATION/hal-01026193

Hurni H, Berhanu D, Gete Z (2011) Sustainable land management in the Ethiopian Highlands. Adaptive Management in Agriculture, Box 9. Highlands and Drylands—mountains, a source of resilience in arid regions. Published by FAO, UNCCD, Mountain Partnership, Swiss Agency for Development and Cooperation (SDC), and CDE, with the support of an international group of experts. Rome, p 76

Lama N (2011) Rangeland management and eco-tourism. In: Kreutzmann H, Yong Y, Tichter J (eds) Pastoralism and rangeland management on the Tibetan Plateau in the context of climate and global change. Deutsche Gesellschaft für Internationale Zusammenarbeit (GIZ) GmbH, pp 87–101

Lama AK, Job H (2014) Protected areas and road development: sustainable development discourses in the Annapurna conservation area, Nepal. Erdkunde 68(4):229–250

Leiva JC (1999) Recent fluctuation of the Argentinean glaciers. Glob Planet Change 22:169–177

Luedeling E, Nagieb M, Wichern F, Brandt M, Deurer M, Buerkert A (2004) Drainage, salt leaching and physico-chemical properties of irrigated man-made terrace soils in a mountain oasis of northern Oman. Geoderma 125(3–4):273–285

Manzardo AE (1977) Ecological constraints on trans-Himalayan trade in Nepal. Contrib Nepal Stud 4:63–81

Miehe G (1984) Vegetationsgrenzen im extremen und multizonalen Hochgebirge (Zentraler Himalaya). Erdkunde 38:268–277

Miehe G, Miehe S, Schlütz F (2009) Early human impact in the forest ecotone of southern High Asia (Hindu Kush, Himalaya). Quatern Res 71:255–265

NTNC [National Trust for Nature Conservation] (2008) Sustainable development plan Mustang (2008–2013). NTNC/GoN/UNEP Kathmandu, Nepal, 102 p

Pandeya B, Buytaert W, Gurung P, Regmi S, Chapagain PS, Bhusal JK, Paudel D (2015) High altitude agricultural practice—the challenge of adapting to the changing water supply in the Himalayas. In: Proceedings Perth III: mountains of our future earth (4–8 Oct 2015, Perth, Scotland)

Paudel KP, Andersen P (2010) Assessing rangeland degradation using multi temporal satellite images and grazing pressure surface model in Upper Mustang, Trans Himalaya, Nepal. Remote Sens Environ 114:1845–1855

Peissel M (1967) Mustang: a lost Tibetan kingdom. Delhi

Pohle P (2000) Rock engravings of Mustang (Nepal)—a unique source for research in cultural history and geography. In: Thapa RP, Baaden J (eds) Nepal—myths & realities, Delhi, pp 521–536

Ponte R (2006) Historia del regadío. Las acequias de Mendoza, Argentina. Scripta Nova, Revista electrónica de geografía y ciencias sociales. Barcelona, vol. X, nº 218 (07)

Poudel DP (2008) Livelihood and common-pool resources. A study of Thini Village, Mustang, Trans-Himalayan Region of Nepal. M.Phil., University of Bergen, Norway, 117 p

Ramirez P (2009) Le Népal entre la Chine et l'Inde. Outre-Terre 1(21):235–241. doi:10.3917/oute.021.02356

Robillard J (2010) L'accès à l'eau dans l'Oasis de Valle de Uco (Province de Mendoza, Argentine). Quand l'agriculture sort des périmètres irrigués traditionnels pour transformer le désert en vignoble. Master thesis, Université Paris I Panthéon Sorbonne, p 99

Sacareau I (1999) Les transformations d'une haute montagne par le tourisme: le massif des Annapurna dans l'Himalaya du Népal/ Tourism and change in a high mountain: the Annapurna area in the Nepalese Himalaya. In: Annales de Géographie, t. 108, 605. pp 21–45. doi:10.3406/geo.1999.21766

Sacareau I (2009) Changes in environmental policy and mountain tourism in Nepal, Revue de Géographie Alpine. J Alpine Res [online], 97–3. doi:10.4000/rga.1018, http://rga.revues.org/1018

Seiter E (sd) Upper Mustang—cultural heritage of Lo Tso Dhun. The Culture Unit, UNESCO Office, Kathmandu, 146 p

Simons A, Schön W (1998) Cave systems and terrace settlements in Mustang, Nepal. Settlement periods from prehistoric times to the present day. Beiträge zur allgemeinen und vergleichenden Archäologie 18:27–47

Simons A, Schön W, Shresta SS (1998) Archaeological research in Mustang. Report on the field work of the years 1994 and 1995 done by the Cologne University team. Ancient Nepal 140:65–83

Snyder DB, Carr WJ (1984) Interpretation of gravity data in a complex volcano-tectonic setting, Southwest Nevada. J Geophys Res 89(B12):10193–10206

Su Y, Lu J, Manandhar S, Ahmad A, Xu J (2013) Policy and institutions in adaptation to climate change. Case study on tree crop diversity in China, Nepal, and Pakistan. ICIMOD working paper 2013/3, Kathmandu: ICIMOD, 60 p

Sun HB, Wang RH, Yang GJ (2007) The Mountain-Oasis-Desert System and Characteristics of Climate in the Arid Zone of Center Asia. A Case Study in the Northern of Xinjiang in China and the East of Kazakhstan. J Arid Land Resour Environ 10:6–11

UNDP [United Nations Development Programme] (2006) Upper Mustang Biodiversity Conservation Project. Final Report of the Nepal UMBCP Terminal Evaluation Mission, 71 p

von der Heide S (2012) Linking routes from the silk road through Nepal—the ancient passage through Mustang and its importance as a Buddhist cultural landscape. In: Archi-cultural translations through the silk road 2nd international conference, Mukogawa Women's University, Nishinomiya, Japan, July 14–16, 2012, vol 613, pp 353–359

von Fürer-Haimendorf C (1975) Himalayan traders. J Murray, London 315 p

Wichern F, Luedeling E, Müller T, Joergensen RJ, Buerkert A (2004) Field measurements of the CO_2 evolution rate under different crops during an irrigation cycle in a mountain oasis of Oman. Appl Soil Ecol 25(1):85–91

Changing Agricultural Practices in the Oases of Southern Tunisia: Conflict and Competition for Resources in a Post-revolutionary and Globalisation Context

Irène Carpentier and Alia Gana

Abstract In Tunisia, the rise of new social demands with regard to regional and local development, particularly in the central and southern areas of Tunisia, has highlighted the conflicting and political dimensions of development strategies. In the oases of southern Tunisia, in particular, changing market conditions and growing competition for land and water, especially between old oases and new productive irrigated areas and between agricultural investors and small farmers, contribute to politicising the issue of access to resources and territorial development. In this context of a changing economic and political environment, we observe a variety of development dynamics in the oasis areas, which results in major transformations in the socio-spatial logics and functioning of these agricultural territories. Based on the field surveys with local actors and farmers in the oases of Tozeur and Gabès, this paper explores the links between new forms of valorising oasis territorial resources and political motivations behind rising demands for social and territorial justice, and the ways in which these dynamics contribute to opening up a new space for public debate about local development models. We formulate the hypothesis that changing management practices of oasis resources in a context of political and eco-environmental crisis, under the pressure of social mobilisation, contributes to new ways of integrating oasis territories into globalisation processes. These are expressed in contradictory and conflicting dynamics. On the one hand, there is a process of de-territorialisation, linked to the development of agribusiness,

I. Carpentier
LADYSS, Paris, France

I. Carpentier
IRMC, Tunis, Tunisia

I. Carpentier (✉)
Panthéon Sorbonne University, 91 rue St Jacques, 75005 Paris, France
e-mail: i.carpentier88@gmail.com

A. Gana
CNRS, IRMC, Tunis (USR 3077), Tunisia

E. Lavie and A. Marshall (eds.), *Oases and Globalisation*,
Springer Geography, DOI 10.1007/978-3-319-50749-1_9

export-oriented intensive farming and mass tourism and, on the other hand, a process of re-territorialisation resulting from innovative practices, based on a heritage valuation of oasis resources, which contribute to rehabilitating diversified family farming systems.

Keywords Sustainability · Changing practices · Tunisia · Ancient oasis · Globalisation

1 Introduction

The Tunisian revolutionary process of 2010–2011 and the democratic transition phase, from 2011 to today, have revealed growing social aspirations for social and spatial justice. In the central and southern regions of the country, people have been increasingly calling for a reorientation of national policies in favour of these long-forgotten territories and for a better valorisation of their resources. Echoing the growing social demands, a new official political discourse has emerged, which focuses on the need to promote the development of rural areas of central and southern Tunisia.

In the oasis of southern Tunisia, in particular, changing market conditions and growing competition for land and water, especially between old oases and new productive irrigated areas and between export-oriented agricultural enterprises and small farmers, contribute to politicising the issue of access to resources and territorial development. The changing economic environment of oasis farm activities also leads to major transformations and differentiation of resource management practices, which result in the reconfiguration of the socio-spatial logics and the functioning of these agricultural territories. These transformations are expressed in contradictory dynamics (Carpentier and Gana 2014). On the one hand, there is a process of de-territorialisation, linked to the development of agribusiness, export-oriented intensive farming and mass tourism and, on the other hand, a process of re-territorialisation resulting from innovative practices, based on the model of heritage valuation of oasis resources. The latter contribute to rehabilitating diversified family farming systems. It should be noted that these dynamics can be integrated into productive strategies at the level of the farm unit. For example, some agrotourism projects, which are fully integrated into world markets, base their marketing strategies on the valuation of territorial resources.

In this context, the purpose of this paper is to explore the dynamics of change that characterise farm management practices and the ways in which these changes modify the socio-territorial constructions associated with oasis farming systems. Our aim is also to examine to what extent the changing economic and political environment, including new forms of integration into globalised markets and networks, opens up new opportunities for the development of innovative and more sustainable models of oasis resource valorisation. Finally, our objective is to assess whether changes in oasis farming systems, as well as territorial dynamics associated with these changes, modify local power relationships or reinforce existing hierarchies.

Our analytical approach rests on the following assumptions:

First, we consider that the transformation of oasis farming is an active process resulting from the socio-economic strategies deployed by farmers and local actors. Very often, the links between oases and global economic processes (modernisation and globalisation) are thought of in terms of the impacts these processes have on oasis farm systems and territories, viewed as passive victims that "suffer" the consequences of these processes (Gana 1998). Rather, we suggest reconsidering this link as producing contradictory processes, which generate both constraints and opportunities, and which simultaneously strengthen and "deny" local territories (Théry 2008). Our approach is thus to take into account both processes of homogenisation and differentiation of oasis farm systems, and their differentiated outcome on oasis territories. In order to identify the socio-spatial dynamics associated with the transformation of oasian farm activities, we use the definition of the territory as "various forms of relationships to space that individuals and social groups continue to produce and process in the context of their social relations" (Alphandéry and Bergues 2004).

This paper is based on the results of interviews and questionnaire surveys conducted with a hundred farmers and with representatives of the local administration in Gabès and Tozeur, between 2011 and 2014. From the data collected, a typology of oasis farming system dynamics was developed, which integrates the elements of the national and global context.

We first analyse the processes and public policies at the origin of the deep crisis of the oasis system, whose consequences are now a threat to the survival of these environments.

Secondly, we look at how this crisis and the political upheavals have led to increased conflicts, claims and mobilisations, which are challenging development models and management policies. We show how the crisis characterising oasis areas since the 1990s is not so much an environmental one as a crisis of management models, which is closely linked to political and economic choices. We also point out that growing social mobilisations express the rise of an associative activism, which plays an increasing role in linking local demands with strategies of global actors.

In the third section, we analyse adaptation strategies of farmers to changing economic conditions and the process of social differentiation of oasian farms.

Finally, we explore the ways in which changing farm practices contributes to the production of a reticular oasian territory, which challenges the traditional boundaries of the oasis.

2 Crisis of the Oasis Farming Systems in a Context of Globalisation

The transformation and crisis of small family farms were extensively documented by rural sociologists and geographers in the 1970s and 1980s (Mendras 1967; Marsden 1991). However, they were mainly analysed as leading to the disappearance of the peasantry or to its progressive integration into capitalist agriculture. This approach tends to overlook the diversity of processes that characterise the transformation of family farming and pay little attention to the links between farmers' practices and structural changes in agriculture (Gana 1998).

In this paper, we aim to analyse the transformations of oasian systems, considering that local farmers' adaptation strategies to changes in their global environment contribute to challenging current development models. In Tunisia, the implementation of structural adjustment policies was translated into new development strategies, which sought to increase the integration of the Tunisian economy into global markets, in particular through the intensification of agricultural exports. In the agricultural sector, including in the oases, structural adjustment policies resulted in the privatisation of the management of state farms, agricultural price liberalisation, the reorganisation of the farm credit system and a reallocation of economic resources to large export-oriented farm enterprises, mainly in the irrigated farm sector, considered to be the most competitive in global markets.

In the oases, the manifestation of the negative impacts of this increased competition on resources between the different productive areas, coupled with the relative failure of economic diversification projects (Abdedaiem 2009; Report Oasis 2012), has reinforced the picture of an oasis faced with a major environmental crisis and characterised by outdated structures and management practices.

2.1 Tozeur and Gabès: Two Examples of the Oasis Crisis

The oasis crisis resulting from modernisation policies and the associated environmental degradation, particularly the overexploitation of water resources and the prospect of losing these historic farmlands, has been a favoured subject for social science research since the 1970s (Dollé and Toutain 1990; Abdedaiem 2009; Kassah 1996; Sghaïer 2010; Bessaoud 2013; Kilani 2014; Baduel and Baduel 1984). Located in the Jerid and Djeffara regions, respectively, Tozeur and Gabès belong to the category of traditional oases (Kassah 1996); Tozeur is a continental oasis, while Gabès is a coastal one (Fig. 1).

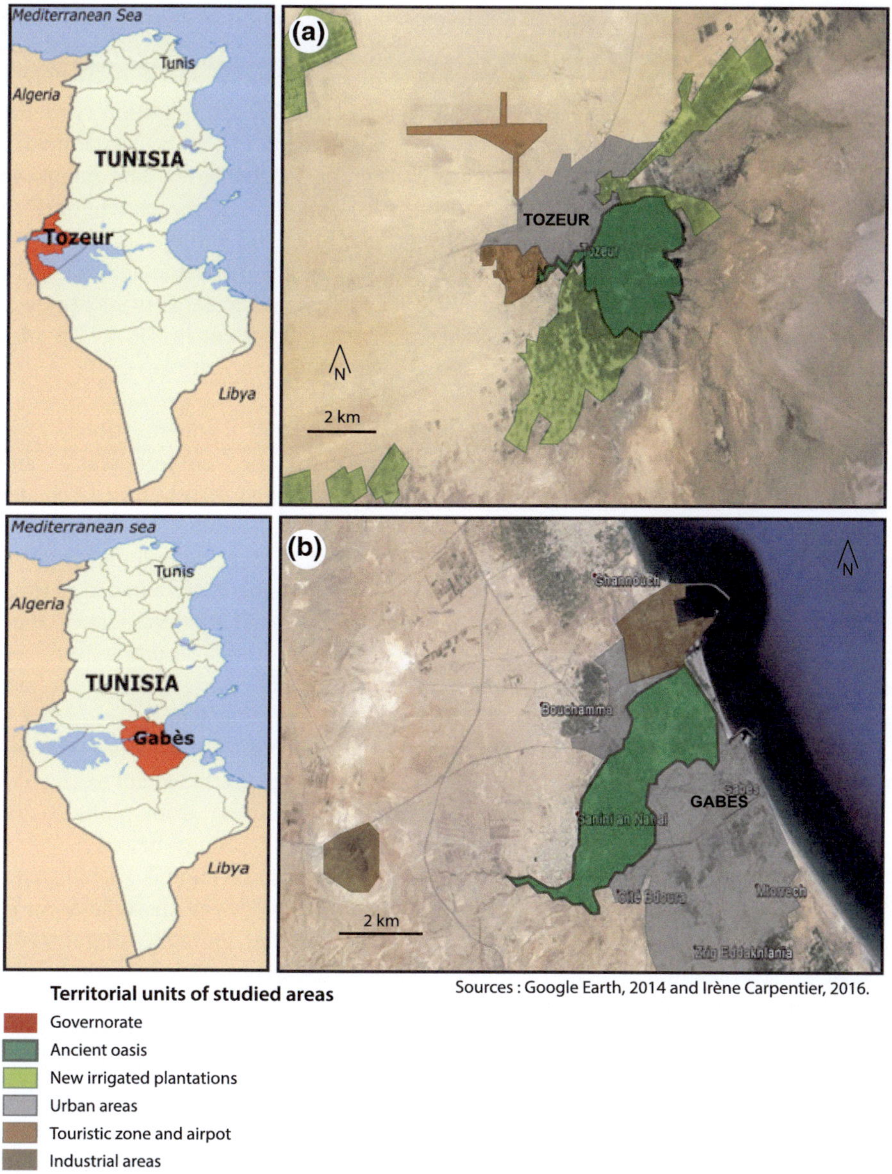

Fig. 1 Tozeur and Gabès, traditional oases in the south of Tunisia

Historical anthroposystems, Global Important Agricultural Heritage Systems (GIAHS[1]) for the FAO, green belts for southern cities, the oases of Tozeur and Gabès symbolise these historic lands, which are often fantasised (Battesti 2005). They also illustrate the specific dynamics of socio-spatial reconfiguration currently underway in the context of changing national politics and the global economy.

The development of new irrigated plantations of Deglet Nour date palms on former state lands, since the colonial era, competes directly with the ancient oasis of Tozeur.[2] The tourist zone, located at the head of the historical springs, as well as the development of new urban districts in the wake of the implementation of the international airport of Tozeur in the 1990s, has reshuffled the cards of local power relationships and local dynamics. In Gabès, the establishment in the 1970s of a chemical industrial complex on the coast and a cement plant upstream of the spring area have generated a similar process.

As a result of these modernisation policies, the oases of southern Tunisia have undergone profound changes that have affected the environmental and socio-economic balance on which the oasis system was historically based (Hames and Kilani 1993; Dollé and Toutain 1990; Lasram 1990; Kassah 1996). While the different territorial units that constituted the ancient oasis were complementary, the oasis territory is now under the pressure of increased competition between various economic activities and territorial units for access to and exploitation of their resources (Fig. 2). Many phenomena, such as urbanisation, desertification and environmental degradation, are putting under threat the oasis traditional farm activities and territories.

The dynamics that characterise the oases of Tozeur and Gabès today are significant examples of the variety of ways in which farming systems adapt to changes in their economic environment, especially to those resulting from globalisation processes. These different strategies are presented in the typology below.

Usually defined as an intensification of flows of goods and capital, globalisation also contributes to integrating local territories into networks of dissemination of models and practices, which go beyond national institutional and policy frameworks (Gana and Terrazzonni 2014). In Tunisia and in the oases, these processes have been translated into the implementation of liberalisation policies (reduced subsidies, higher production costs), creating new conditions for access to land, water resources and markets, thus imposing adaptations that often respond to external demands.

The policy discourse, which after independence presented the oasis areas as "shadow areas" of national development (Battesti 2005), is now developing a new narrative, which underlines the need to protect the oases, considered a national heritage and a major stake in a sustainable national development.[3]

---

[1]GIAHS project, www.fao.org/giahs/

[2]Between 1976 and 1979, irrigated areas in Tozeur grew from 4660 to 7900 ha, a 69% increase (Oasis Report 1999; Ministry of Agriculture).

[3]Strategic report of GIZ and Ministry of Environment, April 2012, "Tunisian oases, a need for protection against degradations and climate change impacts".

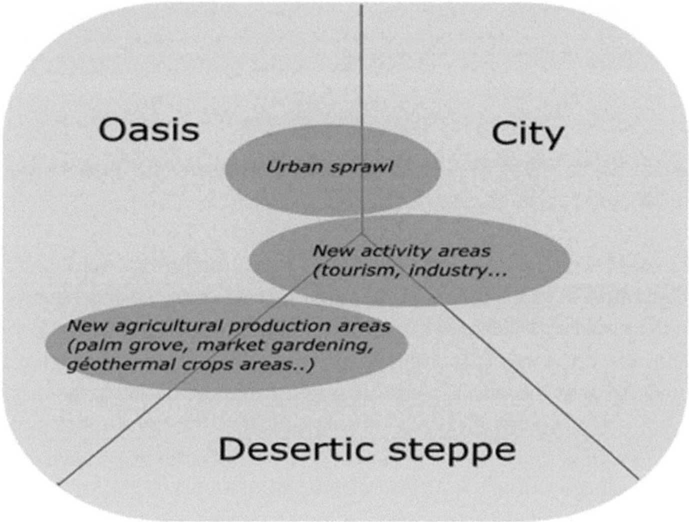

Fig. 2 Diagram of the oasian regions spatial organisation (a traditional trilogy (town, oasis and desert) is now associated with new territories: activity zone and new agricultural production areas)

3 Increasing Conflicts and Demands

3.1 Announcing Conflicts and Social Mobilisations

In the context of the degradation of oasis agrosystems and the crisis of regional development models (Abdedaiem 2009), rising conflicts and social demands for fair access to resources could be observed well before the revolution. Local mobilisations were mostly organised around environmental issues, as this was the only way people could exercise some freedom of expression in an authoritarian context (Battesti 2012). Thus, mobilisations around water problems were often used as an "excuse for political expression", and oasis farmers' discourse tended to develop a general criticism of public policies, but in the framework of local claims (Battesti 2005).

Thus, farmers' criticisms were mainly about constraints to accessing water resources, water being the main "meeting point between the state and market gardeners" (Battesti, *op. cit.*). Poor maintenance of the network and the inequity of the management system, combined with the transfer of the control of the resource from the historic *jamaaia* (community) to Agricultural Development Groups (GDA), contributed to making the issue of access to water a recurring local demand. A farmer of the Tozeur oasis reported that, before the revolution, he had no guaranteed access to water although he was paying for his share, while the chairman of the GDA was able to irrigate his land without paying and to benefit regularly from additional irrigation hours. In fact, the chairman of the GDA very often had close ties with political authorities and could control the allocation of water resources. The interference of influential actors and the mismanagement of water users' associations also weakened smaller farmers' access to water and triggered protest movements at the oasis scale.

3.1.1 Gabès: Social Mobilisations Against Pollution, the Beginning of a Specific Associative Activism

In Gabès as well, most protests were about environmental issues, particularly the pollution generated by the Tunisian Chemical Group, but also land degradation and the negative impacts of urbanisation. Since the colonial era, Ben Salah (2011) estimates that the oasis has lost 30% of its palm trees while the Agricultural Land Agency considers that the oasis is now losing 10 ha per month on average and that, since the revolution, urbanisation has doubled. Before the revolution, statistics were not always available and not very accurate for these sensitive issues. After the revolution, the acceleration of urbanisation, due to the release of political and legislative power, favoured the launch of new studies on these phenomena. The Agricultural Land Agency is now the best source of information to monitor land dynamics.

These dynamics of environmental degradation of the oasis have led to the development of associations whose mission is the protection of the oasis, and this associative movement expanded after the revolution. In fact, the number of associations rose from two to nine, all working for the development of the oasis and the city of Gabès and two working on the specific question of the urbanisation of the oasis. As for the associations that were active before the revolution, we can mention the example of the Chenini UNESCO Club,[4] created in the late 1980s and the Association for the Safeguard of the Oasis of Chenini Gabès (ASOC),[5] established in 1992. The organisation of a seminar on "The impact of pollution on the oasis" by the Chenini UNESCO Club and the Global Movement of Young Ecologists in 1989 resulted in a series of prohibitions by the Governor, who first refused the seminar and then required the presence of the police. The Tunisian Federation of UNESCO Clubs, however, enables the development of activities and mobilisations. In 1992, the Tunisian Federation organised an international meeting entitled "Oasis, heritage of humanity" in Gabès, in partnership with the UNESCO Clubs of France, at which the French agro-environmental activist, Pierre Rabhi, presented a paper on

[4]UNESCO Clubs are an associative movement that supports UNESCO ideals. They also help to spread these in various fields, and carry out activities strongly linked to the principles of the Organisation: "Contributing to peace and security by promoting, through education, science and culture, the collaboration between nations in order to further universal respect for justice, law, human rights and fundamental freedoms for all, without distinction of race, sex, language or religion, as the Charter of the United Nations recognises to all the peoples" (Constitutive Act, Charter of UNESCO).

[5]ASOC: *Association de Sauvegarde de l'Oasis de Chenini Gabès* (Association for the Safeguard of the Gabès Chenini Oasis).

"Peoples' food autonomy". Following this meeting, the now oldest and most established associative structure of the oasis of Gabès was created, the Association for the Safeguard of the Oasis of Chenini (ASOC). However, under political pressure, the UNESCO Club had to stop its activities. The ASOC focused initially on a project aimed at protecting the soil from environmental degradation, with the introduction of agro-ecological practices, such as home composting, conducted with "pilot" farmers. Subsequently, its activities were diversified with the creation of a Maghrebian Network of Associations for Sustainable Development in the Oases (RADDO[6]) in 2001, and the building of a national collective development project for the oasis territories within the ISPO[7] programme (Initiatives for Preservation of Oasis Heritage) linking all oasian associations in Tunisia. These projects are deployed in a regional context dominated by actors of the chemical industry, which makes it difficult to question the responsibility of the chemical group for the environmental degradation of the oasis. Claims for a "right to health", a safeguarded environment, and the maintenance of a viable oasian territory are mobilising people beyond the strict sphere of oasis farmers. These activists are emerging without open conflicts with local institutions and central government, using a priori non-political repertoires, such as environmental and heritage protection. This enables local actors to build a critical discourse, which is actually quite radical. Since the revolution, these associations have been criticised for choosing to work with the administration.

3.1.2 Tozeur: Mobilisations for a New Tourism Model, More Respectful of the Ancient Oasis

In Tozeur, the rapid development of mass tourism since the 1990s, in connection with the scarcity of oasis resources, gave way to an early contestation of regional resource management policies and tourism development, increasingly perceived as "outrageous" by local actors. The Tozeur UNESCO Club was created as a result of a meeting between the Regional Culture Delegate of Tozeur, a native of Gabès, and the President of the Gabès UNESCO Club. It began working for the rehabilitation of the oasis and mitigation of the tourism impact on it in the late 1990s. Since 2008, the Nafta UNESCO Club, with the support of international donors, has offered an ecotourism tour in the oasis of Nafta. Over a course of 20 hectares, this tour involves 206 owners around the old river bed, which is now concrete and replenished by drilling, and part of a "visual re-enchantment of the oasis" (Battesti 2005). Tourism, with the water issue, remains the economic sector that polarises local mobilisations and dynamics. The Association for the Safeguard of the Medina of Tozeur (ASMT), founded in 2001, was for a long time the only symbol of associative action in Tozeur. It tries to promote the development of a new form of

[6]RADDO: *Réseau des Association de Développement Durable dans les Oasis.*

[7]ISPO: *Initiatives de Sauvegarde du Patrimoine Oasien.*

tourism, more respectful of environmental constraints, more "ecological and cultural".[8] An educational display system in the medina, hiking trips in the oasis and raising citizens' awareness of the negative impacts of mass tourism on the local territory are the main projects of this association. Without being in open conflict with the administration or private actors, these associative groups contribute to the emergence of a local discourse calling for the reorientation of development policies, especially in the agricultural and tourism sectors.

3.2 Post-revolution Mobilisations, the Crisis of Resource Management Policies and Their Political Implications

The Tunisian revolution significantly expanded the scope of these social and associative dynamics and discourses. Oasis sustainability and promotion has become a political issue at the regional scale. At the local level, claims have moved from the narrative of sustainability to that of social and spatial justice, echoing the opening of the national political arena and debate.

We analyse here two examples of "places" of crystallisation of post-revolutionary mobilisations: Agricultural Development Groups (GDAs), and cooperatives and farmers' unions. These were chosen because they are representative of the movement of contestation of development models and of the politicisation of issues related to resource management, which favoured the renewal of the debate on territorial development. Social protests are now much more confrontational, mobilising the most marginalised actors among farmers. The functioning of these organisations illustrates the type of interactions between farmers and institutions. For a long time, they also symbolised the fossilisation and corruption characterising the management of resources at a local level.

It is important to recall that local debates on resource management models take place in the larger framework of the post-revolution debate on the development of southern Tunisia (oil, gas and phosphate)[9] and participate in the critique of governmental action and policy orientations.

3.2.1 The Agricultural Development Groups (GDAs): Management Crisis and Political Expression for Fair Governance

The GDAs are the associations in charge of the distribution of water for irrigation in the oasis. In Tozeur, there are four active GDAs for four *jars* (oasian neighbourhoods) of the oasis: Hafir, Abbes, Wasat and R'bat. In Gabès, there are also four

[8]www.tozeurasm.org.

[9]As an example, we could mention the "*winou el petrol?*" (Where is the oil?) movement. For more information, see: http://www.huffpostmaghreb.com/2015/05/27/winou-el-petrole-tunisie_n_7449592.html.

GDAs: Chenini, Chott El Ferik, Menzel and Jara. These associations, which include all irrigators in the oases, crystallise conflicts over management models and access to resources. Although the general distribution principle is perceived as fair by the majority of farmers (2014 surveys), criticisms of the functioning of these management structures are widespread: lack of human and financial resources, financial opacity and inefficiency in maintaining water distribution networks… In fact, these management structures, created after the policy of state withdrawal, are one of the symbols of the fossilisation of local resource management systems. In Tozeur, GDAs are indebted for an average sum of 50,000 dt (25,000 euros) to the STEG[10] and 15,000 dt to the CRDA,[11] due mainly to political corruption, non-payment, and a lack of support from the administration for maintaining the drilling. The chairman of the GDA of Hafir also mentioned the fact that the billing system to the STEG is not adapted to a seasonal agriculture as it is a monthly payment. All oasian GDAs are today facing structural problems. The renewal of the leaders, accused of corruption, has not eliminated the crisis of governance. The rejuvenation of the GDA boards after the revolution was often accompanied by a political struggle for the control of these strategic structures. Moreover, especially in Tozeur, outdated drilling, blackouts and poor irrigation water quality have pushed irrigators to refuse to pay for their water shares, since GDA debts were not reduced. The discourse of some local actors stresses how farmers do not trust GDAs, and asserts that social injustice, which has long characterised water management institutions, is not really challenged. Small farmers are still the most vulnerable to shortages. GDA general assemblies, expected to occur once every two years, are rarely held. In Gabès, the director of the Chenini GDA has been replaced four times since 2011 but this has not been successful in resolving the governance crisis.

Contestation of GDA functioning takes place in a context of growing struggles for financial transparency and equity, and against corruption of the distribution system. Open conflicts sometimes emerge between water users and managers, GDAs and the STEG, and GDAs and the CRDA. GDAs have become favoured spaces for a political expression aimed at redefining management models.

3.2.2 Agricultural Cooperatives and Unions: Increasing Conflicts and New Political Implications

All the same, cooperatives and farmers' unions are challenged for their dysfunction, which results in more conflicts and contestations. These structures are denounced for being symbols of the monopoly of marketing channels. They are clearly identified by farmers as responsible for restricting better resource valuation opportunities for producers. In Gabès, there is only one milk collection cooperative, which operates for all the breeders of the oasis (23% of farmers). Therefore, it has

[10]STEG: *Société Tunisienne de l'Électricité et du Gaz.*

[11]CRDA: *Commissariat Régional de Développement Agricole.*

control over the purchase price and terms of sale. This weakens the small farmers who are dependent on the conditions imposed by the cooperative. It is the same with the UTAP, the largest union of Tunisian farmers that controls the marketing of concentrated animal feed. Claims are not politically structured but in Gabès, farmers have tried to organise themselves to ensure the marketing of their products and to escape this pressure on prices and conditions of distribution. Furthermore, the expansion of the Libyan market opened new informal markets. A representative of the animal production service at CRDA declared that over one-third of the milk produced in Gabès is marketed in Libya and does not go through the health monitoring service.[12]

In Tozeur, the election for the renewal of the Regional Farmers' Union Office in 2012 led to a major conflict. Elections were not recognised as valid by some members, who could not present their candidacy. The presence of political lists for union elections, such as the election of a leader who is not a farmer, made things worse (2012 surveys). The Tozeur Regional Farmers' Union headquarters was even burnt in May 2012, expressing the deepening of conflicts.

These collective organisation structures, which, under the rule of the former regime, were little involved in the promotion of innovative models of oasis resource management and were relatively disconnected from the dynamics of the local territory, are now at the centre of a politicisation of issues related to the management and governance of local resources.

4 Oasian Dynamics in a Post-revolutionary Context: Differentiation of Opportunities and Strategies

These dynamics of crises and conflicts have resulted in major transformations of agricultural activity and contributed to the emergence of new adaptation strategies. The political discourse in favour of promoting a new development model takes place in a context of growing social demands and changing farm management strategies.

The analysis of the changing management practices of the "traditional" oasis territory enables us to characterise the diversity of agricultural adaptations in a context of deep economic and political changes. Local actors (civil society actors, private investors, small farmers, etc.) are the protagonists of differentiated development strategies: technical modernisation of conventional methods, adaptations of the traditional model and alternative models of heritage valorisation (Carpentier and Gana 2014).

In the following section, we examine the variety of ways in which farmers have been adapting to changes in their economic environment and how the Tunisian revolution is likely to have had specific impacts on those dynamics. These impacts,

[12]Field surveys at the animal production service at the CRDA, 2013.

particularly on social territorial dynamics, are of course not clearly visible, as the process of political change is still underway.

Let us first also stress the fact that the precise counting of each farm type is not an easy task, as there is no agricultural census in Tunisia. The following table (Table 1) is therefore built on the criteria we used to elaborate the typology. Based on the typology and the data collected at the level of the GDA regional administration, we have tried to extrapolate the proportion of each category of farm adaptation.

4.1 The Model of Oasis Agribusiness

Among this first group of oasis farms, productive strategies and innovations are based on large investments and aimed at consolidating large farm enterprises integrated into national and international markets. Characterised by very modern infrastructures, including electricity connection, wells, ponds, irrigation canals and agricultural buildings, their economic projects aim to conquer export niche markets in particular. Access to water resources is usually ensured by the presence of a well or a valve on the private plot for the autonomous management of irrigation. Crop specialisation is at the heart of the farming system. In Tozeur, this has led to almost a monoculture of the Deglet Nour date. The farms of this category are quite large (over 1.5 ha) and located mainly along communication axes that cross the ancient oasis. This type of farm is the standard in all new plantations. Agricultural profitability of such projects is important, but often part of a pluriactive economic strategy. Agriculture remains a complementary activity for entrepreneurs mainly active in other sectors and places, such as Tunis or Sfax, bigger cities than Tozeur. Nevertheless, it enables those who are native to the region to maintain a link with the local territory and ensure membership of the oasis community. This type represents approximately 13% of the oasian farmers (in Tozeur) (based on the number of farmers who have more than 3 ha) and 38% of the cultivated area of the oasis. Among them, 9 farmers manage their commercialisation themselves and 38% have a private hydraulic infrastructure. The biggest farmer in Tozeur, also a date trader, is known to irrigate his plots with one-third of the oasis water. In Gabès, it is very rare to find such entrepreneurs (0.01%) but there are a few vegetable traders who manage the biggest plots.

These agribusiness projects are fully rooted in a logic of growth based on integration into export markets and international client networks. Their productive systems and landscape models are characterised by the removal of the layering of crops and alignment of palm tree plantations. Representing a minority of oasis farmers, the promoters of these projects control almost two-thirds of the cultivated area of the ancient oasis and contribute to accelerating land concentration. More generally, they take advantage of a dominant position in the competition for access to local resources. For example, as mentioned above, one farmer alone uses one-third of all the irrigation water of the Tozeur oasis.

Table 1 Forms of adaptation to globalisation in the oases of Tozeur and Gabès

Typology of forms of adaptation to globalisation				
Type of valorisation	% of oasis farmers	% of the cultivated areas—spatial impact	% of the water consumption	Goals of each type of valorisation
Agribusiness	13 (Tozeur)	38 (Tozeur)	38 (Tozeur)	Economic profitability, agricultural productivity
	0.01 (Gabès)	3 (Gabès)	3 (Gabès)	
Territorial marketing	18 (Tozeur): 2% of the farmers	2–3 ha on average (Tozeur): 4.4%	4.4	Aesthetic, leisure, landscape valorisation
	12 (Gabès): 0.4% of the farmers	1–2 ha on average (Gabès): 2.2%	2.2	
Diversified and pluriactive family farming	70% of the farmers (Tozeur)	35.1	35.1	Food, family, heritage, social territory
	63% of the farmers (Gabès)	49.2	49.2	
Specialised and intensive family farms	10% of the farmers (Tozeur)	2.5 ha on average (22.5% of the oasis territory)	22.5	Productivity, specialisation, economic profitability
	23% of the farmers (cattle breeding in Gabès)	0.6 ha on average (20% of the oasis territory)	20	
Oasian activists	5 associations in Tozeur	Walking tour in the oasis		Creating new development models and collective dynamics
	137 farmers (of around 600 in Chenini; 22.8% of Chenini farmers; 4% of all the farmers of Gabès oasis) have a convention for associative projects in Chenini Gabès (organic farming and local seeds); 5 associations	60 ha of organic farming in Chenini Gabès (8% of the oasis territory)	8	
Abandoned plots	No reliable information	No reliable information	(No irrigation)	End of cultivation, urbanisation, desertification
	15%	17.25%		

These agro-entrepreneurs often also claim their contribution to the sustainability of the oasis territory and to the promotion of its heritage, which they reduce to the date palm. They point out that their activities can prevent land abandonment and preserve the environment and its productive function. These narratives, which present the oasis agricultural activity as a local economic resource that can ensure environmental sustainability, are part of regional projects aimed at promoting models of territorial valorisation based on market logics. Quality labels or conversion to organic agriculture of recent date palm tree plantations in the governorate of Tozeur make it possible for farmers of these areas to seize new opportunities, and position themselves in the "heritage" and "alternative products" markets, responding mostly to international demand. These strategies are promoted and supported by development institutions. In Tozeur, the Ministry of Agriculture has developed a plan for the conversion of all oasis plantations to organic farming by 2030, with the aim of renegotiating the export markets, and optimising the profitability of Deglet Nour dates.[13]

The revolution seems to have had little impact on this category of farmers, at least from a socio-economic point of view. Their ability to promote a specialised farming system is based on their insertion into and mastering of national and international marketing channels, and also on their relative autonomy regarding access to resources. However, they complain about rising labour costs. The daily wage has doubled since the revolution, and this has become a problem for these agro-entrepreneurs whose farming activities are based on the use of hired labour or the traditional oasis sharecropping, the *khamessat*. The owner of a big farm expresses his concerns: "Who will climb the palm trees? These workers are acrobats, and there are no more acrobats. Maybe in five years it will be a disaster". Another one points out: "They have become very lazy since the revolution, and ask for more money. The workforce of the mining area doesn't come anymore, just because it is no longer profitable". On the other hand, *khamess* and workers are trying to organise themselves in order to renegotiate their working and safety conditions. Significant changes in the work organisation in oasis farming are underway and are likely to lead to a renegotiation of local power relationships.

4.2 Territorial Marketing, a Tertiarisation of Oasian Activity?

The second group of oasis farms can be characterised as taking part in a tertiarisation of agricultural activity, which appears to be a result of a new development policy orientation. We define it as the model of "territorial marketing", as the main activity is the supply of services, such as recreation and tourism. These activities are a means to promote the "oasian scenery" responding to the demand of national and international tourists. However, in the current context of the tourism crisis, these recreational and restoration areas are increasingly used by local people and entrepreneurs trying to adapt to this situation. This is the case of the "Sahara Lounge" in Tozeur, a tree-climbing area on a plot of R'bat, which also offers a cafe and

[13]2011 surveys.

entertainment space. Various activities are available on the same plot, in order to attract a wider clientele: restaurants, cafe, games for children, a party room and film projections. Other examples are "El Berka Park" in Tozeur and "Amine Park" in Chenini. The productive function of agriculture is marginal and is preserved just for the direct use of customers and staff (especially fruit trees, used for lemonade, orange juice and dates). The oasis area is no longer considered a food production area but a place of leisure and "marketing". The referent landscape is that of the traditional oasis. It is maintained and recreated by gardening, in which the aesthetic dimension overpowers the others. Decorative plants, sometimes sterile and imported, are cultivated. In fact, the added value of the territory is achieved by the marketing of a historic and diverse landscape, exploited for leisure. The oasian entrepreneurs are in the process of opening up oasian territories to service activities. The diversification of oasian activities that meets international and local demand is a way for the local actors, who control most of the local economy, to seize new opportunities in the context of an increasingly tight market, in both agriculture and tourism.

The shock of the revolution hit the tourism sector severely, particularly in Tozeur, where nearly half of the hotels have closed. The promoters of polyvalent oasis leisure activities, which represent the latest generation of tourism projects within the oasis, took advantage of this crisis. The development of a family leisure activity in the oasis is a significant revolution in the *Jerid* region, where previously only the palm workers were present (Battesti 2009). In Gabès, the development of these leisure areas responds to the demands of the urban youth of Gabès city, usually more attracted by the sea and the "corniche" for their leisure activities, and finding in the oasis a renewed social space. In Tozeur, these oasis leisure areas generally attract a local and urban clientele.

4.3 Diversified and Pluriactive Family Farming

Alongside these new oasian entrepreneurs, "peasant" forms of valorisation of oasis resources persist, drawing on crop diversification strategies. With a low investment capacity, small farmers deploy their projects within the family farm unit. They represent around two-thirds of oasian farmers, but less than half of the cultivated land of the oasis, as they control very small plots. With a highly fragmented territory, the operation of the land property by small farmers varies significantly according to access to water, accessibility of the plots and soil salinity. Cropping systems are diverse and multi-layered; they are predominantly oriented towards fruit tree production or horticulture according to the local agricultural zoning; farms often have an associated small livestock activity. In the practice of the activity, as in the arrangement of plots and crop management, the "traditional" model is dominant despite the growing influence of the modernist model, under the pressure of the market. The inherited skills (shrub transplants, changes in irrigation practices and crop association) are the basis for adapting to the constraints affecting this category of farms, in particular lack of water, restricted access to land due to the competition of industrial activities and urbanisation, and marketing problems. Self-consumption, especially of vegetables and fruits, remains one of the main functions of the productive activity. The crops are not always economically profitable but contribute to household food security. In Tozeur, it is the agricultural labour force, the *khammes*, who maintain the layering of cultures and the consumption function of the plots.

However, for small farmers to maintain their activity and stay on their land, further integration into the market is required. In Gabès, small landowners have developed an economy based on pluriactivity and the promotion of the oasis territory in order to adapt to new market conditions and opportunities. In fact, the arrival of Libyan migrants, who have specific consumption demands and habits, has led to the introduction of new crops, such as leeks and aubergines, which where traditionally absent in the oases. This is a way for small farmers to differentiate themselves in a limited and saturated local market, and adapt to water stress. Thus, market gardeners, who have to deal with the increasingly fierce competition from other productive areas, have developed a speculative agriculture. The practices of these small producers preserve the historical functions of the oasis: landscape function, feeding function and socio-economic function. The objective is to ensure the reproduction of the system and the farm unit, while preserving and promoting family heritage.

These small oasis farmers constitute the majority of the ancient oasis farmers but control land rarely exceeding 0.5 ha (0.12 ha in Gabès on average, with 90% of farmers who own 0.5 ha; 1.02 ha in Tozeur, with 59% with less than 1 ha[14]). Economically and socially marginalised, this group of farmers is most exposed to disruption linked to the revolution. For example, two- or threefold increases in the prices of phytosanitary products or the daily wages of the labour force are often difficult constraints to overcome for these small farmers. Security issues for herds and water infrastructures also make this category particularly fragile, in a context marked by political instability and the overall weakening of the government.

4.4 Specialised and Intensive Family Farms

Unlike the previous category, a group of small- and medium-sized farmers deploy strategies of intensification and specialisation, fostered by state-supported programmes, for the development of cattle rearing and market gardening projects (Gabès). They benefit from the stability of market prices and provide their promoters with regular income throughout the year.

Among this group of farmers, productive logics and resource valorisation are based on strategies of intensification, specialisation and integration into regional markets. In general, the oasis landscape tends to be conserved, particularly through

[14]According to GDA statistics (2013 surveys).

the date palm, which plays the role of preserving the ecosystem. In some cases, however, fruit tree plantations have almost disappeared and have been replaced by forage crops in order to feed cattle.

Economic projects are based on medium-sized investments that are encouraged by state-supported programmes. The hydraulic infrastructure (irrigation channels) is well maintained by the farmers themselves to optimise available resources. Cattle producers benefit from a supplementary valve that ensures water as required. They pay each year to water users' associations. The whole plot is devoted to the main business, cattle rearing or market gardening. For cattle projects, alfalfa is grown on most of the land, whereas for market gardening projects all plots are dedicated to vegetable production.

The size of this category of farms is between 0.3 and 0.6 ha and each one of the plots that constitutes the often-fragmented land property is dedicated to a specific crop according to its situation. In the oasis of Gabès, in Menzel and Chenini, the introduction of cattle breeding programmes has encouraged this type of farmer. They are not found in Tozeur. The agricultural profitability of such projects is not always very large, because of the sharp increase in prices for livestock feed. Only the biggest are profitable. Most farmers in this category are pluriactive. Being very specialised, this group of farms is heavily dependent on market prices, the quality of state support and marketing networks. After the revolution, these farmers became particularly involved in the contestation of the cooperatives and farmers' unions and in the debate about technical models of farm development and marketing-related problems.

4.5 The Oasian Activists

The political and ideological dimension of development choices is expressed in the emergence of a "new" category of farmers, the "oasis activists". They are the promoters of an alternative development model, based on the rehabilitation of the oasis environment through the use of traditional skills and the promotion of agro-ecological practices. Calling into question the prevailing modernist and liberal development model, the "alternative" movement emerged long before the revolution, but was reinforced by the new political context.

This type of valorisation of oasis resources is based on the diversification of economic activities: farm production, farm product transformation and ecotourism. Promoters of "alternative" practices are members of associative networks, which seek to provide small farmers with answers and advice to solve the problems they face at a local level: soil quality, water availability, diseases, etc. The "oasian activists" are part-time farmers. They have a specific sociological profile. They are often teachers, employees, civil servants, or trade union representatives and belong mostly to the upper middle class. Their projects are deployed in diverse collective networks. Associations for the protection of oases bring them together in Tozeur and Gabès. Since the revolution, the proliferation of environmental protection and local development associations has contributed to a greater diversity of people involved in these networks.[15] The opening of the political space and the enrichment of the associative environment have also diversified issues of debate and negotiation, from land-related issues to the organisation of cultural events, and actions aimed at strengthening the autonomy of smaller producers.

With a medium-sized investment capacity and production unit, this group of oasian "activists" practises farming on their own land using family labour and sometimes a limited number of workers. Available land is cultivated according to the traditional landscape model. Cropping systems are diversified and layered, from the palm tree, to forage, industrial and vegetable crops. Crop association, rotation and fallow techniques are promoted. Traditional skills are at the heart of the valorisation strategies, such as seed management and irrigation. These farms do not differ specifically from other forms of peasant family farming. In fact, their own practices matter less than the formation of a collective project for territorial development. From this point of view, the revolution has helped associations to reconnect with a political discourse promoting these forgotten areas of national development.

The conquest of autonomy in the management of resources is a key issue for association projects dealing with environmental problems, and induces a radical critique of situations of monopoly with regard to the commercialisation of dates and milk in Tozeur and Gabès as well as the malfunction of GDAs. The construction of an alternative model is based on the transformation of oasis practices and functions. The nourishing function of the oasian market garden is replaced by an educational one, which is the basic unit for a new development model. These actors are trying to generate, through associative projects, a new local and urban demand, sensitive to better product quality and the safeguard of the environment. Better control of prices by operators, establishment of an organic farm sector, and farm product transformation activities are the main goals around which the network for an alternative model of heritage valorisation is organised. The projects are deployed locally and sometimes via international associative networks, for example the export project of organic pomegranates from the oasis of Gabès, managed by ASOC.[16] In the long run, the aim is to develop service activities related to tourism, culture and handicrafts in the oasis, which would increase the multifunctionality of the oasian

[15]*Association de Sauvegarde de la Medina de Tozeur* (ASMT), Club UNESCO of Tozeur, Post-revolution: *Association de Sauvegarde de l'oasis de Tozeur* (ASOT) created in 2011, *Association des Amis du Palmier de Tozeur* created in 2011; *Association de Sauvegarde de l'Oasis de Chenini* (ASOC) created in 1992, Post-revolution: *Association Formes et Couleurs Oasiennes* (AFCO) created in 2011, *Gabès Action* created in 2011, *SOS Gabès Environnement* created in 2011.

[16]In 2014, 107 plots and 52 ha of the oasis of Chenini in Gabès were approved for organic farming by a convention with ASOC. Furthermore, in 2013, 178 farmers were given free fruit trees by the association to promote biodiversity in the market gardens.

territory. Alternative tourism is a key element of these diversification strategies. Educational farms, farm accommodation, eco-accommodation and organic catering are some associative initiatives redefining oasian tourism. These initiatives are viewed as likely to preserve cultural resources and the landscape of the oasis, in accordance with international "cultural" tourist expectations (Battesti 2013). The "traditional" oasis model is mobilised through its most symbolic areas: Ras el Oued in Chenini and the periphery of the Ouled el Hej district in Tozeur. The oasian associative actors aim to challenge the modernist model and propose innovations as part of the transformation of the traditional model. The sustainability of this alternative oasis model relies on the diversification of agricultural activities, considered the condition for preserving the integrity of the environment and the landscape. However, the ability of this model to counteract dynamics characterised by the disappearance of the oasian environment seems strongly linked to government support, which since the revolution appears more willing to open the debate on development models.

A process of differentiation of oasis farming systems and social groups can be observed, as a result of the following dynamics: economic marginalisation of small farms, which increasingly limit their activity to family food consumption objectives, specialisation of a group of large enterprises in leisure and tourism-related activities, intensification and development of cash-generating crops for a better integration into marketing channels. In Tunisia's post-revolutionary context, changes in local development dynamics are increasing social disparities and modifying local power relationships. In fact, the period of political upheaval and social unrest aggravated the crisis and the constraints for farmers, such as the availability and access to water and farm inputs, and increased market and price instability, resulting in unequal opportunities for farmers to valorise available resources. In this context, contradictory trends are observed with regard to territorial dynamics associated with the transformation of farming systems and the restructuring of oasian agriculture: increased land fragmentation, farm land concentration, processes of de-territorialisation, linked to the development of export-oriented date palm tree production, reinforcement of the territorial anchoring of agricultural activity resulting from the rise of innovative practices drawing on the valorisation of the oasis heritage.

In the following section, we explore further these territorial dynamics.

5 Territorial Dynamics and Socio-spatial Disparity

According to Théry (2008), the "territory" is more than a bare support comprising inventoried and saleable resources; it is primarily an asset for the actors who practise it and try to promote it, playing more or less successfully, the great globalisation game. In the oases, the development of the monoculture of Deglet Nour has strong impacts on the landscape and cultural implications (geometric alignment of the trees, disappearance of the lower crop level and intensification of irrigation). These dynamics result from strategies deployed by the group of oasis enterprises in order to seize new market opportunities and take advantage of the high prices of palm tree dates, especially the organic ones. Seeking a better positioning in the global competition, these strategies consolidate specialised areas within the oasis. In fact, the international date market has expanded over the past ten years with the emergence of new highly intensive and competitive production areas in the Emirates and North America. As a result, producers in the traditional oases are faced with fierce competition. In this context, the reactivation of the historical productive function of the ancient oasis is viewed as a means of being more competitive in the international date market. Nevertheless, the local territory dimension is not promoted enough in the marketing strategy. The dates of Tozeur are not labelled and marketed as ancient oasis dates, "products of oasian heritage", but as "conventional" products of the Jerid region. Conversion to organic farming in Tozeur is not really conceived as a way of promoting a new model of production, but rather as a way of competing better in the global date market, thanks to a quality-based strategy.

However, the opening of oasian territories to worldwide competition, the emergence of a new demand for "rooted products", and recent dynamics of promotion of "cultural" tourism by local stakeholders and the Tunisian government have increased this quest for "authenticity" and "local", making the local territory an object of transaction (Battesti 2009). Some oasian actors have responded to this demand for "roots" to "sell" their territory better through developing projects emphasising the specificity of the oasis.

On the other hand, the opening of oasian territories to globalisation dynamics has contributed to the dissemination at a transnational scale of ideas and practices, such as the conceptualisation of a comprehensive and universal ecological issue, and the need to protect what belongs to a world heritage. For example, the objective of the FAO GIAHS programme is to label the Saharan oasis as a world heritage. In the same spirit, since the late 1990s, local actors and associations of Gabès have been working towards the registration of the Gabès oasis on the list of "Treasures of Humanity". Local development actors in the oasis of Gabès are taking advantage of this movement, especially since the Tunisian revolution, and are seizing new funding opportunities provided by international institutions and donors. The creation of a label of oasian products (for example, pomegranates in Gabès), the establishment of collective or cooperative organisations and the development of eco-tourism projects illustrate some of the ways in which local actors in the oasis of southern Tunisia are diffusing "alternative" models of territorial resource valorisation, as well as being involved in the socio-spatial restructuring of the oasian territory.

Another illustration of how social actors and oasian farmers are reconfiguring the oasian territory is the "network of *khamess*" in Tozeur (Fig. 3). Regardless of the mesh of the plot, the *khammes* (sharecroppers), as members of a collective network, contribute, through their practices, to producing a specific territory. Each *khamess* is in close connection with his neighbour and farm work is coordinated and collectively organised. This coordination standardises choices of farm resource valorisation and creates a homogeneous landscape. Another example is provided by farm workers in the area of Abbess. Originating from the village of Sahraoui, they organised themselves to create a specific cultivated area, integrating date palm trees

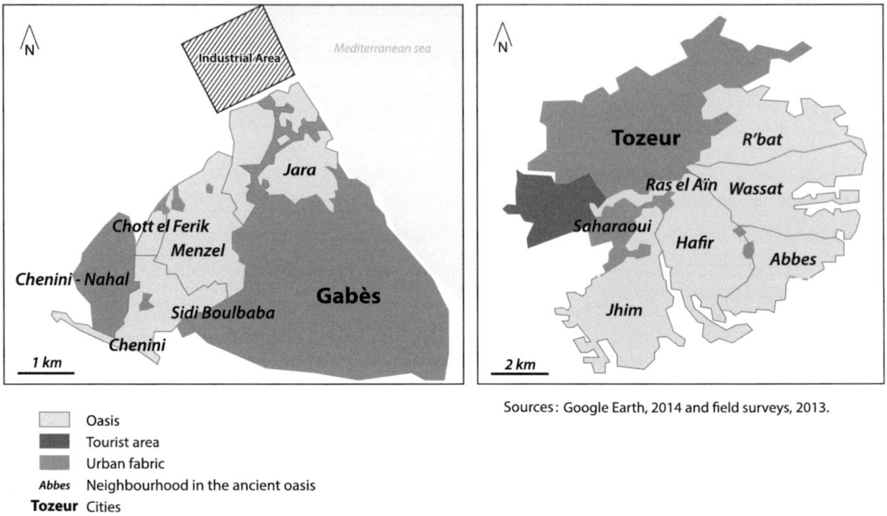

Fig. 3 Territorial organisation of Tozeur, and Gabès, an oasis in the town

and fig trees, but at the expense of pomegranates. They also invested in ground crops, such as mallow, parsley and condiments.

These strategies, which have various motivations, lead to the construction of an oasis micro-territory. Since the revolution, the emergence of a collective claim for the suppression of the social status of the *khamessat*, who have no right to land, no legal recognition or social security, has drawn on these networking spaces. Despite conflicts and difficulties, these dynamics are imposing a dialogue between these farm workers and land owners. A *khamess* expresses these difficulties: "We have to organise ourselves in workers' unions, so far we haven't been able to talk with our employers, they make it harder on us". This mobilisation of *khamess* for their social recognition is one of the most striking elements of the contestation of existing power and work relationships since the revolution.

In Gabès also, the construction of a territory as a space of action is particularly visible in Chenini. The oasis of Gabès surrounds the city and the socio-spatial logics in the oasis reproduce those of the neighbourhoods from where the landowners originate (Fig. 3).

Thus, the Chmessa area, north-east of the oasis of Chenini, is "disconnected" from the rest of the oasian territory. The owners, originating from the neighbourhood of Sidi Boulbaba, do not belong to the same territory of action. For example, associative actors of Chenini do not deploy their action in this area, even though it is in the municipality of Chenini. Similarly, the neighbourhoods of Menzal and Jara, which were historically in conflict, deploy differentiated strategies of resource valorisation.

Hence, social inequalities and spatial dynamics contribute to the construction of differentiated micro-territories, which are more or less connected to networks and main axes enabling the fulfilment of development potential.

6 Conclusion

In connection with the rise in local demands calling for the redefinition of development models and strategies, the issue of the sustainability of the oases of southern Tunisia has become a political stake. So far, analyses of oasis problems have focused primarily on issues related to environmental sustainability and the protection of the oasis heritage, overlooking the social and political stakes. Since 2011, the deployment of local initiatives in the framework of collective organisation has highlighted claims for social justice and equal access to development opportunities in the oasian areas.

The opening and integration of the oases of southern Tunisia into new markets, while enabling the emergence of new actors and the development of new activities, has not resulted in a process of standardisation, but rather an increased differentiation of oasis systems and territories. In the post-revolutionary national context, the dynamism and diversity of local initiatives for the promotion of traditional oasis territories go against the fatalistic idea of the ineluctable decline of the oases, which has long prevailed in these spaces. At first, the revolutionary episode seemed to have reinforced the dominant actors and existing trends in the oases. In reality, the rise and diversification of social demands and mobilisations for a more equitable development model have contributed to a more comprehensive approach of the oases "crisis", which considers their sustainability not only in environmental and agricultural terms, but also from a social and political point of view. Being historically crossroads territories, the oases are privileged spaces for this analysis of how local actors participate, through their initiatives and projects, in processes of social and economic change and in making a new development model, likely to answer the demands of an oasian population eager to carry out its revolution.

References

Abdedaiem S (2009) Socio-spatial changes and water modes of governance in suburban oasis of Gabes governorate (Southeast Tunisia). Thesis (PhD) in Geography, Paris 10 University

Alphandéry P, Bergues M (2004) Territoires en questions: pratiques des lieux, usages d'un mot. Ethnologie française 34(1):5–12

Baduel AF, Baduel P-R (1984) Une oasis continentale du Sud-Tunisien. In: Revue de l'Occident musulman et de la Méditerranée, N°38, pp 153–170

Battesti V (2005) Jardins au désert: évolution des pratiques et savoirs oasiens, Jérid tunisien, IRD, Coll. A travers champs, p 440

Battesti V (2009) Tourisme d'oasis. Les mirages naturels et culturels d'une rencontre? In: Cahiers d'Etudes africaines, 2009/1-2 (n°193–194), Tourismes, la quête de soi par la pratique des autres, EHESS, pp 551–582

Battesti V (2012) The power of a disappearance: water in the Jerid region of Tunisia. In: Johnston BR, Hiwasaki L, Klaver IJ, Ramos Castillo A, Strang V (eds) Water, cultural diversity and global environmental change: emerging trends, sustainable futures? UNESCO/Springer, pp 77–96

Battesti V (2013) Des ressources et des appropriations. Retour, après la révolution, dans les oasis du Jérid (Tunisie). In: Etudes rurale, EHESS, 2013/2, (192), pp 153–175, <hal-00832566v2>

Ben Salah M (2011) La palmeraie de Gabès, Phoenix project, France-Italy

Bessaoud O (2013) Le grand chambardement des oasis maghrébine, Actes de colloque. In: Bathaoui G (ed) Terres et vies oasiennes face au défi du changement climatique. RADDO, Gabès, Tunisie

Carpentier I, Gana A (2014) Les oasis de Tozeur et Chenini Gabès: durabilité et diversité des formes de valorisation à l'ère de la mondialisation et des crises du développement. In: Marshall A, Lavie E, Chaléard J-L, Fort M, Lombard J (eds) Proceedings of the colloquium: oases in globalization: ruptures and continuities, Paris, 16–17 Dec 2013, pp 105–112

Dollé V, Toutain G (eds) (1990) Les systèmes agricoles oasiens. CIHEAM, Montpellier. Options Méditerranéennes, Série A, Séminaires Méditerranéens; n°11

Gana A (1998) Agricultural restructuring, household practices and family farm differentiation, a case study of the region of Zaghouan. Thesis (PhD), Cornell University, 287 p

Gana A, Terrazzonni L (2014) Quelles contributions de la sociologie et de l'anthropologie à l'analyse de l'intégration régionale. In: Gana A, Richard Y (eds) La régionalisation du monde, construction territoriale et articulation global/local. IRMC Khartala, 272 p

Hames C, Kilani M (1993) La Construction de la mémoire. Le lignage et la sainteté dans l'oasis d'El Ksar. In: Archives de sciences sociales des religions. N. 82, pp 284–285

Kassah A (1996) Les oasis tunisiennes. Aménagement hydro-agricole et développement en zone aride. In: Série géographique, n°13. CERES, Tunis, 346 p

Kilani M (2014) The collapse of the oasis under the impact of the state. The Tunisian case. In: Colloque international: Les espaces oasiens: des écosystèmes à la mise en valeur patrimoniale et touristique, Paris Sorbonne University of Abu Dhabi (Reem Island), Abu Dhabi

Lasram M (1990) Les systèmes agricoles oasiens dans le Sud de la Tunisie. In: Dollé V, Toutain G (eds) Les systèmes agricoles oasiens. CIHEAM, Montpellier, pp 21–27. (Options Méditerranéennes: Série A. Séminaires Méditerranéens; n. 11). Les Systèmes Agricoles Oasiens, 1988/11/19-21, Tozeur (Tunisia)

Marsden T (1991) Theoretical issues in the continuity of petty commodity production. In: Whatmore S, Lowe P, Marsden T (eds) Rural enterprise: shifting perspectives on small-scale production. David Fulton, London

Mendras H (1967) La fin des paysans, Paris, SEDEIS, Coll. Futuribles, 361 p

Ministry of Agriculture (1999) Oasis report, Tunis

Sghaïer M (2010) Etude de la gouvernance des ressources naturelles dans les oasis. Cas des oasis en tunisie, UICN, 69 p

Théry H (2008) Mondialisation, déterritorialisation, reterritorialisation (globalization and territory). Bulletin de l'Association des géographes français, 85ème année, 2008-3 (September). Mondialisation et géographie/L'ouest américain, pp 324–331

Tourism Development in the Thar Desert, Rajasthan

Philippe Cadène

Abstract Crossed by rich caravans at the time of the silk roads, the Thar cities suffered a series of crises from the middle of the nineteenth century to recent decades. But since the 1980s, the ancient desert capitals have experienced a new process of growth, tourism being the main development sector. This chapter analyses the process by which this activity started and has expanded to our days. It gives a description of the rich heritage of these cities, with their forts, temples, and havelis, the merchant caste's magnificent houses. It shows the importance of foreign visitors' arrival, the active role of the former nobilities in providing adequate accommodations, the use of strategic infrastructure for tourism development during the 1980s and 1990s. After 2000, globalisation brings drastic changes. The opening of Indian economy leads to external investments in quality hotel infrastructure. The rise of a national middle class expands domestic tourism. If expertise, investments, and political decisions come from outside the desert cities, local factors still play a large role in such remote regions, even in the case of tourism, an activity driven by an external demand.

Keywords Tourism · Desert · Oasis · Regional development · Urban development · Globalisation

Crossed by rich caravans at the time of the silk roads, the Thar Desert suffered a series of crises from the middle of the nineteenth century to recent decades. Controlling trade routes, the splendid desert cities were specifically affected by the modernisation of sea transport and the building of a colonial economy organised from a few ports established by the British (Kulke and Rothermund 1990). Rich merchants left for the large cities of colonial India, mainly Bombay and Calcutta and contributed to their development. From 1947, the partition of the colonial

P. Cadène (✉)
UMR CESSMA, University Paris-Diderot, Sorbonne-Paris-Cité, Paris, France
e-mail: philippe.cadene@univ-paris-diderot.fr

P. Cadène
Université Paris-Diderot, 5 rue T. Mann, CC7001, 75205 Paris Cedex 13, France

E. Lavie and A. Marshall (eds.), *Oases and Globalisation*,
Springer Geography, DOI 10.1007/978-3-319-50749-1_10

Empire between Pakistan and India deepened the crisis. A hard border suddenly separated the desert into two parts and the remaining trade collapsed. Except for military cantonments, the only development of the desert cities after Independence was linked with the building of a modern administration. The crisis also reached the villages, mainly inhabited by shepherds. The land was being degraded by grazing and logging, while the still growing population was increasingly engaged in a constant search for water.

However, since the 1980s, the Thar cities have experienced a new process of growth. In the eastern part of the desert, Jodhpur, the largest of them, with 1,138,300 inhabitants according to the 2011 census, has become one of the major tourist attractions in north-west India. In the central-west part of the state, close to the Pakistan border, Jaisalmer, with only 60,000–70,000 people, has been intensely visited for the beauty of its fort, palaces, and other buildings (Sharma 2012). In the north-western part of the area, also near the border, Bikaner, with around 650,000 inhabitants, started to attract visitors only at the end of the 1990s.

Tourism is indeed the faster growing sector in these three desert cities (Fig. 1), located far from the highway crossing Rajasthan from north to south, the Delhi–Mumbai NH 8. This study analyses the process by which this activity started and has expanded to this day. It aims to demonstrate how local factors still play a large role in such remote regions, even in the case of tourism, an activity driven by an external demand. Expertise, investments, and political decisions come from outside the desert cities. This is the case for hotel management, investment in tourism infrastructure, and the building of strategic roads in the desert to address the tensions with Pakistan. However, the resources and their valorisation, embedded in the local urban societies, appear essential for explaining the change over the last three decades. The development process of the Thar cities relies on an interaction between external and local forces in the context of globalisation.

1 The Rich Heritage of the Ancient Desert Capitals

The Thar cities certainly offer some of the most impressive heritage areas in India due to the large number of monuments as well as the richness and diversity of their architecture. The heritage buildings can be differentiated according to their traditional functions. The most impressive, both in size and spatial extent, are the forts and palaces built by the former Rajput rulers. The main temples, most of them rich in fine carvings, are the centres of religious life in the cities. Today, most of them belong to the state, but they continue to be the symbolic places where the Brahmin class stands. The *havelis*, often huge mansions decorated with paintings and marble or wooden sculptures, were built mainly by rich members of merchant communities and still belong to these groups of families, who continue to control a large part of the local economy (Cadène 2000).

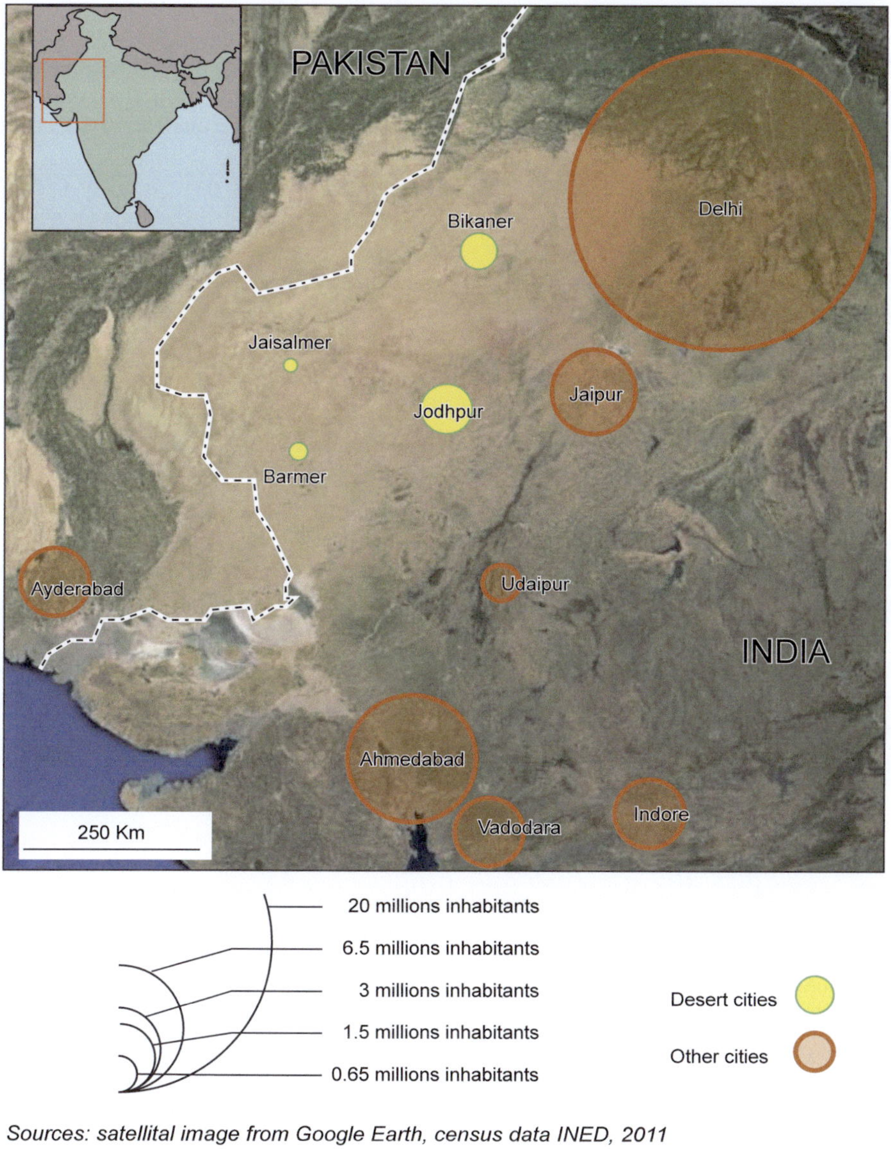

Fig. 1 Thar desert and its cities in north-west India

1.1 The Forts and Palaces of the Rajput Nobility

These three cities, Jodhpur, Jaisalmer, and Bikaner, were the capitals of three former princely states. They were part of British India but ruled by their royal families, who retained their status until the end of British rule (Fox 1971). As the process of Independence came along with the integration of the "princely states" into the modern Indian state, kings and nobles lost their political power (Stern 1991).

Although some of the nobles had to give up part of their heritage to the state, the families who built the forts and palaces themselves could keep a significant proportion of them, where they were often living. Their impressive architecture, along with the ancient hospitality of the nobles and their political influence, enabled these places to become the main tourist spots in the cities from the British period onwards (Tillotson and Rupert 1987; Sharma 2008; Taft 2003).

Jodhpur offers the most perfect example of this situation and the importance played by the nobility heritage in tourism development. An impressive fort dominates the capital of the former kingdom. South of the historical centre, an amazing palace, built in the 1930s combining Indian and western architecture, has been converted into a luxury hotel. The cities of Bikaner and Jaisalmer also offer their visitors beautiful forts, magnificent palaces and several other buildings related to the history of the local nobility. This is also the case in many smaller towns, even if they were not important seats of power. Mandawa may be cited, with its fort that was also turned into a hotel in the mid-1990s. However, other forts and palaces, equally impressive, still remain outside the main circuits, as in Nagaur or Khimsar.

1.2 Temples and the Brahmin Elite

India is famous for the complexity of Hindu polytheism, the religion of about 80% of its population. There are innumerable temples in towns and villages, where they are part of the built landscape and anchored to the practice of the Hindu religion. Their size varies, largely depending on the status of their main deity, the status of the communities that built these places of worship and the status of the priests who serve the gods. According to the Hindu system, each large temple is a world in itself, populated by priests who belong to Brahmin castes and organise religious life. Still today, they alone have the right to enter the sacrosanct place and to perform rituals close to the gods. However, members of other castes historically contributed to temple construction and continue to ensure their operation through donations. The most beautiful temples were often built by members of the royal or merchant castes, who continue to participate actively in their management and provide a significant proportion of devotees (Reiniche 1990). This is particularly true in Rajasthan, where the power of kings and rich merchants contributed to the erection of the most famous temples.

Many holy Hindu places are today well known in the Thar Desert. They are usually visited by tourists, Indians and foreigners, on their way to the cities. Not far from Bikaner, in the small town of Deshnok, the Karni Mata Temple, famous for its sacred rats that are revered, attracts pilgrims and tourists. Further east, at the edge of the desert, Pushkar, a small town where one of the few temples dedicated to Vishnu is located, is visited by a large number of foreign tourists, attracted by the mystical atmosphere of the place. In the south, a few kilometres from Mount Abu, a tourist resort that dominates the desert, visitors are amazed by the Jain temples of Dilwara, the whiteness of their marble walls, and the quality of the sculptures that adorn them.

However, many of these temples are not regular pilgrimage centres and remain unvisited by tourists. Multiple explanations can be given for this lack of interest. A major reason lies in the management of these religious sites. The main objectives of the boards of directors or communities who run the temples are to ensure the perpetuation of rituals, the maintenance of the building, the management of donations, and the organisation of pilgrimages. Their aim is certainly not to transform the temples into tourist attractions. These heritage places are valued for their deities more than for their architectural beauty.

1.3 The Merchant Caste's Havelis

With the *havelis*, the Rajasthani merchant castes also have their architectural treasures. These luxurious mansions generally date from the nineteenth century or the beginning of the twentieth century. Merchant families built them for their business as well as for their residence (Timberg 1978). *Havelis* can be found in cities and in small market towns. They are large buildings, closed to the outside and sometimes fortified. A rivalry developed between merchant families to impress their clients, which led to an escalation in both the size of the buildings and the richness of their decoration (Cooper 2009).

In the western part of the desert, the *haveli* windows are particularly sophisticated with fine balconies enclosed by *moucharabiehs* or, in contrast, graceful oculi provided in the upper floors, all designed to capture the slightest breath of air when the heat wave prevails in the warm season.

Long before Independence, many *havelis* were deserted by their owners, who left for other regions of India. The situation of these buildings varies from city to city. They are rarely well maintained, even in tourist areas. Their superb murals are particularly badly damaged. Although they attract tourists, their absentee owners do not worry about the future of the architectural wealth they possess, especially the younger generations. Only some of them have been transformed into heritage hotels or museums. Some can be visited if the watchman employed by the owning family authorises tourists to enter in exchange for a little money.

2 1980s–1990s: The First Steps of Tourism Development

Except for a few British officers of the Colonial army or administration, and some rare visitors, the first westerners to visit the Thar cities were the hippies travelling to India in the late 1960s. They opened the way to a large number of foreigners in search of exoticism (Lagadec 2003). In the late 1970s, when the former kings and noble families started to understand the potential of tourism development, they rapidly decided to transform their heritage buildings into hotels. They targeted foreign visitors according to the prestige of their palaces and mansions. Independently, transport infrastructures were built, largely for geostrategic reasons. Large roads opened up the desert and led to the arrival of more tourists, accompanied by the development and diversification of the tourism business.

2.1 *From Hippies to Backpackers*

The splendour of the Thar Desert cities was known to some curious and rich visitors in the nineteenth century. However, their opening up to modern tourism was due to young western travellers who visited these places during the 1960s. They followed a movement that arose within the most developed countries, particularly the USA, which promoted a radical critique of the consumer society established in the west in the middle of the twentieth century. Their choice of India was explained by the image of peace and mysticism often attributed to this country, the freedom of movement within it, the tolerance of lifestyles, and, certainly, the possibility of obtaining various drugs in some areas. They selected some places, where they lived on a tiny budget. The Thar Desert was on their itinerary. At the edge of the desert, the pilgrimage centre of Pushkar, in particular, attracted these young travellers with its mystical atmosphere. From this small town, they could explore the desert cities. Thus, Jodhpur and Jaisalmer appeared very early on the Indian tourist map, although, at this time, their impressive fortresses and palaces were difficult to access.

In fact, the hippie movement fabricated a fantasy of India in western countries (Lagadec 2003). It created a taste for India among members of the North American and European middle classes who, in the following decades, flew to Delhi or Bombay and developed a new geography of tourism in India.

In the 1980s, the cities of the Thar Desert were listed among the main places to visit on the international tourist maps. The hippies were replaced by other western young people, less idealistic; the backpackers. They were budget tourists on extended or working holidays, avoiding the most crowded destinations (Alder 1985). Still difficult to reach, but relatively cheap, the Thar cities offered perfect destinations. Year after year, the backpackers arrived in their thousands.

2.2 The Arrival of Wealthy Westerners

The Thar Desert started to welcome a growing number of international tourists from the end of the 1980s. Although many were backpackers, travelling with little money, with each year the economic status of the tourists increased. Many of them were affluent. Holding university degrees, occupying well-paid jobs, they were attracted by the cultural richness of the cities. They often reached Jaipur or Udaipur by plane, then rented taxis, took buses or sometimes trains to travel towards the desert cities.

Arriving in the desert, these foreign tourists met an Indian society built specifically for them by the local stakeholders of the tourism economy: hotel and restaurant owners, shopkeepers, guides, and taxi and rickshaw drivers. They visited both artificial and real sites, rooted in ancient communities as well as constructed by the outside. They appreciated staying in heritage hotels. They loved walking in areas retaining an authentic atmosphere and shopping in bazaars where local crafts were sold. Nevertheless, a large part of their stay was spent on the terraces of their hotels, made to satisfy their desire for exoticism (Bautes 2004; Bautes and Valette 2004; Ramusack 1998).

2.3 The Former Maharajas and the Western Elite

With the arrival of these foreign visitors, the elite of the desert cities understood the large potential for tourism offered by their region. However, for such development to take place, the various places had to provide infrastructures of international standard. Notwithstanding, former Maharajas and nobilities had the means to build an appropriate strategy. The kings and their vassals had certainly lost their power after Independence, but they had kept most of their palaces and staff and much of their prestige. For a time, they even enjoyed a civil list provided by the central government, allowing most of them to maintain a high lifestyle. The removal of such pensions in 1971 by Indira Gandhi, the then Prime Minister, led them to search for an alternative. The arrival of tourists offered them an opportunity. The richest decided to transform part of their splendid palaces into prestigious hotels and facilitate the arrival of wealthy tourists.

The Maharajas of Jaipur and Udaipur led the process for tourism development in the entire region. These two historic cities offer an exceptional architectural heritage. At the edge of the desert, both of them benefit from a good connection with Delhi and Mumbai, due to their location on the National Highway No. 8 linking the two major Indian cities. The two Maharajas were the first to understand how the beauty of the Rajasthan royal capitals could promote the development of their state. They engaged in the transformation of some of their palaces into magnificent hotels. In Jaipur, the first to be opened as a luxury hotel was the Rambagh palace, located in the south of the city, where the Maharaja's family lived from the early twentieth century to Independence. In Udaipur, the first hotel was the Lake Palace, the former summer palace of the king's family, located in the middle of the lake. The new hotel was promoted in the media by Jackie Kennedy, a friend of the royal family. It rapidly emerged as one of the top tourist attractions in India (Bautes 2004). These two hotels were soon managed by the Taj Hotels Resorts and Palaces Company, the hospitality branch of the Tata Group that owns the prestigious Taj Mahal Palace in Mumbai.

Fairly quickly, other noble families followed this example. Tourism development in the Thar Desert began with Jodhpur, which benefits from its central location and the beauty of its fort and palace. Closer to the Pakistani border, Jaisalmer began to receive visitors amazed by the beauty of the citadel and the *havelis* located in the bazaar. Finally, during the last ten years, the city of Bikaner, in the north-west of the desert, has begun to attract visitors. On the way, the towns of Shekavati, especially Mandawa, have started to be on the tourist map (Sharma 1999).

2.4 The Diversification of Tourist Activities

Everywhere in Rajasthan, tourism development followed the same path. First, the noble families tried to manage themselves the hotels they had opened in their forts and palaces. Then, unable to provide international standards, they handed over the management to professional companies specialising in the hospitality business. Some of them sold their properties to the Rajasthan state or the National Ministry of Tourism. They were transformed into public sector hotels, mostly receiving groups of foreign tourists on tours. These hotels rapidly became famous, but few of them were able to reach the required international standards.

The number of hotels created in heritage buildings expanded dramatically from the end of the 1980s, when the local lords, the Thakurs, started to follow the example of the Maharajas. They changed their fortresses located in small towns into hotels, as long as they were not located too far from the main roads. They also offered accommodation in the mansions their families had built in the kingdom capitals at the end of the nineteenth century or beginning of the twentieth.

A few years later, merchant families, and any other families able to offer quality accommodation, also engaged in the hotel business. The quality provided depended upon the capacity of the family to understand and anticipate the needs of western clients. Finally, other investors went into tourist activities. In the 1990s, restaurants and shops dedicated to foreign customers multiplied. Small tourism agencies, taxi companies, and shops offering phones and Internet connections were created. A full range of tourist services was offered in the Thar cities and other places in the desert.

In the main cities, another important phenomenon was the return of families who had left decades earlier. Many were members of merchant communities who had played a key role in these places at the time of the caravans. Most of them still owned properties and wished to enhance the family's wealth by taking advantage of the tourism business. Many stayed and worked in the cities only during the peak tourist season. However, their involvement in the urban economy led to a revival of the cities.

2.5 *The Building of a Strategic Transport Infrastructure*

In the 1970s and 1980s, transport in India was slow and people wishing to visit the region had to devote a significant part of their stay to travelling. It was a clear obstacle for tourism development. The change came paradoxically from the tension between India and Pakistan, giving a strategic dimension to this border territory.

Trains were the traditional means of transport for travellers in the desert. Railways connected all the Thar cities from the end of the nineteenth century. It was the best solution for tourists willing to visit these places in the 1970s and 1980s. However, journeys were slow and not very comfortable. The creation of a special train for rich tourists in 1982, "The Palace on Wheels", offering luxurious tourist packages from city to city, cannot be taken into account, being a tourist attraction and not really a means of transport.

Airports did not offer a real alternative. Although there were small airports in the main desert cities before Independence, they were built by the princely states and used by the local elite. After 1947 and with two wars fought in the desert, the Indian army developed military airports in Jodhpur, Jaisalmer, and Bikaner. From the 1950s, a civil area was in operation in Jodhpur, but it received only a few flights a day. It offered some visitors a gateway to the Thar Desert but was not a solution to visiting different places. For decades, discussions took place about the opening of civil areas in Jaisalmer and Bikaner military airports. Recently, projects have been developed, and some investments have been made. The different stakeholders in the tourism business have exercised pressure for the implementation of these projects, but no decision has yet been taken to open the routes by the national airport authority. Negotiations seem to be still going on with the military.

The solution for tourism development came from the roads. Many were enlarged or built in the desert for strategic raisons. At the end of the 1980s, wide and well-maintained roads began to facilitate a rapid penetration towards Bikaner, Jaisalmer, and even Barmer. These cities close to the border, and also the cities of Shekawati on the way from Jaipur to Bikaner, were no longer isolated; they were henceforth accessible in less than a day's drive. Tourism rapidly benefited from these new infrastructures. From the 1990s, bus trips offered a fast and cheap solution for travelling while taxis became increasingly fashionable among high- and middle-class travellers. In the 2000s, car rental also became possible, but few tourists adopt this solution. It is too expensive and unusual for Indian tourists and looks too perilous for foreigners who are not used to the Indian traffic.

3 After 2000: The Time of Globalisation

The recent period has seen the integration of the Thar Desert into the globalisation process. As India is among the fastest-growing economies in the world (Rothermund 2009), tourism investments are coming to the desert from the large Indian cities and from abroad, while the number of visitors is growing. Indian tourists now constitute the bulk of the Rajasthani tourism business, often combining traditional forms of travelling with contemporary tourism. This change is transforming tourist activities, increasing the number of places to visit and opening the desert to urban travellers.

3.1 External Investments in Quality Hotel Infrastructure

The decade after 2000 appeared to be a turning point in the development of tourism in the desert. Large tour operators, some coming from abroad, needed good hotel infrastructures. The palaces owned by nobles and the small accommodation managed by local families did not provide rooms in sufficient numbers. In both places, international standards often remained out of reach. New investments were necessary.

The newcomers followed the model already set up in Jaipur and Udaipur. In these two cities, new luxurious hotels were built and new palaces were opened, all following international standards. Old buildings were completely redesigned for the hospitality business, and new hotels were constructed with old style architecture. In the three largest cities of the desert, Jodhpur, Jaisalmer and Bikaner, some new palaces were built. For example, in Jodhpur, the prestigious Indian Taj Group established the Taj Vivanta Hari Mahal in 2000. This hotel immediately became one of the best places to stay in town. Similarly, in Jaisalmer, the Suryagarhm luxury hotel was built in the surrounding desert on the model of an ancient palace by the MRS Group, a Rajasthani company operating in the marble business and hospitality. The group manages the Laxmi Niwas Palace in Bikaner, the former residence of the Maharaja, built in 1992, and built the Suryagarhm in the 2000s with the help of the Jaipur-based architect Ravi Gupta, who is known for his heritage architecture and who designed the Oberoi Rajvilas in Jaipur. However, the international chains are still rare. One example is the Fern Residency in Jodhpur, owned by the Chaudhary Group, a multi-national conglomerate with headquarters in Kathmandu, Nepal.

One after the other, the Thar cities received four- or five-star hotels for rich visitors. In some cases, these hotels were built in the desert itself, at some distance from the cities. They even offered romantic and comfortable tents to the tourists, such as the resorts of the Rao Bikaji Group, owned by a scion of the Bikaner royal family.

Hotels were also needed to accommodate average-income tourists. Many middle-range hotels were built, specifically on the outskirts of cities. Sometimes these constructions included palace architecture elements, largely expected by foreign tourists. On the contrary, a modern architecture was sometimes used, often preferred by Indian tourists as well as by business travellers.

Finally, simple houses were transformed into guest houses, allowing tourists on a budget to find lodgings that met their needs. Many of these cheap places received young foreigners and used their experience of more than two decades in tourism to provide appropriate solutions. Increasingly, Indian young people also began to travel and stay in these small inexpensive hotels.

3.2 Domestic Visitors Versus Foreigners

The growth of national tourism is an important phenomenon in India, which started during the 1990s. Indians have always travelled for pilgrimages, family rituals or business but, from a population of 846,387,888 people in the Census of India 1991, only 66,670,303 domestic tourist visits were counted by the Indian Ministry of Tourism in the same year. With 1285 million people estimated in 2015, the figure has jumped to nearly 1.5 billion tourist visits (including day excursions). Tourism is thus now a common activity in India (Bautes et al. 2015). Among these domestic tourists, nearly 14% still said they were visiting a temple or going on a pilgrimage, around 60% were travelling for social or family reasons and 8% for business. Nevertheless, about 31 million Indians were travelling for leisure at this date, four times more than the foreign tourists in the country. Moreover, a number of Indians who have to travel for social or religious reasons clearly take advantage of their trips for some sightseeing. The number of foreign tourists in the entire country was only multiplied by four during the same period, going from 1,677,508 in 1991 to 6,967,601 in 2013. India is not indeed a major place for international tourism (Ministry of Tourism 2013).

In Rajasthan, there were 30,298,150 domestic tourists in 2013 and 28,611,831 in 2012, while this figure was only 5,263,121 in 1992 and 902,769 in 1972. There were 1,437,162 foreign tourists in 2013 and 1,451,370 in 2012, while there were 547,802 in 1992 and 48,350 in 1972. The proportion of foreigners in the total number of tourists went from 5.1% in 1972 to 9.4% in 1992 before falling to 4.8% then 4.2% in 2012 and 2013. In this state, the most visited tourist places are pilgrimage centres: Ajmer, the most important Muslim pilgrimage site in India with the tomb of the Sufi saint Moinuddin Chishti (3,753,260 domestic tourists in 2012, but only 30,750 foreigners), Pushkar, an important place of Hindu pilgrimage (2,323,050 Indians, 70,766 foreigners), Mount Abu, a tourist resort on a hill known for its Jain temples (2,312,448 Indians, 11,386 foreigners), Ranakpur, a splendid Jain temple (616,172 Indians, 110,094 foreigners), and Nathdwara, famous for its Krishna temple (616,095 Indians, only 8 foreigners). Only Jaipur, the state capital, with 998,703 domestic tourists and 534,256 foreigners, managed to intrude among these religious tourism sites so highly prized in India (Department of Tourism Rajasthan 2013).

Among these places, Ajmer, Pushkar and Mount Abu are at the edge of the desert. Further to reach, the Thar cities have had relatively limited success with Indian tourists, while foreigners have made the effort to travel to them. Jodhpur received only 383,357 Indian tourists and 121,034 foreigners, Bikaner 324,988 and 76,497, and Jaisalmer only 126,490 and 73,299, respectively.

3.3 *The Opening of the Desert*

The arrival of Indian tourists has expanded tourism patterns in Rajasthan. Among them, religious tourism appears the most important. It highlights specific locations, some of them being famous pilgrimage places, while others have been developed with the arrival of these new kinds of pilgrims, who include religious practices in their leisure trips.

Tourism has also developed in small towns. Most of them were isolated for a long time, and some are still far from the main tourist centres. Yet, places like Barmer, Khimsar, Nagaur, Osiyan, Pokaran, and Rohet now offer some comfort for visitors. They have built hotels and restaurants, and they propose activities to visitors. They try to attract tourists by organising small festivals inspired by those found in the bigger cities. For example, Pokaran holds a festival on the theme of the desert while Barmer, further away, organises the “Thar Festival” every year, which has put the place on the tourist map. These small towns are visited by more and more Indian tourists and have started to attract foreigners.

The desert itself has also become an area opened up for tourism. Tourists leave the cities to discover this unusual environment. “Desert camps” and long-distance trips by car or camel are organised to provide a memorable experience to visitors, whether Indians or foreigners.

The diversification of tourists is matched by the diversification of tourism companies. Some Indian tourism agencies have developed expertise in organising stays in the desert and are essential intermediaries for most of the agencies that offer tourist services to their clients from other Indian states or from abroad. Indian and foreign tour operators are involved in the sale of tourist circuits and stays. Agencies working on the Internet are on the increase, some of them offering lower prices. An intense competition is developing between professionals.

4 Conclusion

The Thar Desert cities play a vital role in the development of tourism in northern India. Jodhpur, Jaisalmer, and Bikaner are very popular destinations for foreign visitors. These cities are also famous among Indian tourists who are now the majority. From these tourist centres, many other places are now visited; small and medium-sized towns and villages, as well as the desert itself where camps are organised, offering visitors the opportunity to go for a four-wheel drive or a camel ride.

Local nobilities began the process in the 1970s and 1980s, highlighting their exceptional familial heritage. Today, these families continue to control many tourism activities in the entire area, but the business has rapidly expanded with the influx of members of the higher castes, who have invested in hotels, restaurants, antique and craft shops, travel agencies, and taxi companies. The members of the lower castes, representing mostly the working classes in these highly hierarchical societies, have also participated in this process: artisans engaged in making handicrafts for tourists, people from the camel herder community becoming scooter-taxi drivers.

The rapid development of tourism activities shows the dynamism of these local societies. Different groups of people are involved in this business, which reproduces and consolidates the social hierarchy. However, all local stakeholders follow an economic model initiated in the two major tourist places of Rajasthan, Jaipur, and Udaipur, outside the desert area, and most of them have difficulty in providing services of international standard.

It is therefore remarkable that a large proportion of the factors involved in the tourism business in the desert cities originates from outside. Specialised companies and qualified people as well as large investments come from the Indian cities, which appear to be the essential sources for the development of modern activities. Although international hospitality companies are still rare, there are many large national hotel groups. This is also the case for the major tourist companies, selling the desert cities from afar and organising tourist tours. Important projects in roads and other public infrastructures are planned by the central government, largely for strategic reasons in this desert crossed by the border with Pakistan.

However, difficulties and obstructions also come from outside. The policy of the central state appears to be ambiguous regarding tourism development. On the one hand, the Indian Ministry of Tourism launched a multiannual international marketing campaign in 2002 to promote tourism in India, which was very successful and won several awards from professional organisations worldwide. On the other hand, the government has difficulty in facilitating visa formalities in its consulates abroad. The different institutions in charge of modernising and maintaining infrastructures struggle to provide adequate services. In spite of investments in Jaisalmer and Bikaner airports, only Jodhpur airport offers some flights to tourists and businesspeople. Trains continue to be slow and largely uncomfortable. Roads are not always maintained. In spite of good telecommunications, booking is still difficult, reinforcing problems due to poor hotel management.

Finally, the cities of the Thar Desert are in urgent need of a comprehensive development strategy to integrate tourism into the regional economy, not at the scale of the desert or at the Rajasthan state level, but at the level of the large area rising rapidly between the two largest cities of India, Delhi and Mumbai.

References

Alder J (1985) Youth on the road: reflections on the history of tramping. Ann Tourism Res 12 (3):335–354

Bautes N (2004) Le goût de l'héritage. Processus de production d'un territoire touristique: Udaipur en Inde du Nord (Rajasthan). PhD Thesis, University Paris Diderot

Bautes N, Valette E (2004) Painting production, cultural economy and territorial dynamics in Rajasthan. In: Scott AJ, Power D (eds) The cultural industries and the production of culture. Routledge, London, pp 207–223

Bautes N, Cadène Ph, Dumortier B (2015) Le tourisme en Inde: imaginaire et réalités. In: Cadène Ph, Dumortier B (eds) L'Inde, une géographie. Colin, Paris

Cadène Ph (2000) Le développement du tourisme au Rajasthan. Usage ou abandon d'un patrimoine fragmenté. In: Montaut A (ed) Le Rajasthan. Ses dieux, ses héros, son peuple. Hommage au Dr. N. Joshi. Edition Langues'O, Paris

Cooper I (2009) The painted towns of Shekhawati. Prakash Books, New Delhi

Department of Tourism Rajasthan (2013) Annual report 2012–2013

Fox R (1971) Kin, clan, raja, rule: state-hinterland relations in preindustrial India. University of California Press, Berkeley

Kulke H, Rothermund D (1990) A history of India. Routledge, London

Lagadec Ph (2003) "Du "pèlerinage aux sources" à la "route des Zindes": Pratiques et représentations", Association Jeunes Etudes Indiennes, Séminaire Jeunes Chercheurs (21 novembre), Poitiers

Ministry of Tourism (2013) India tourism statistics

Ramusack B (1998) The Indian princes as fantasy. In: Breckenridge C (ed) Consuming modernity. Public culture in a South Asian World. University of Minnesota Press, Minneapolis

Reiniche M-L (1990) Tiruvannamalai: un lieu śivaïte du sud de l'Inde. iv: La configuration sociologique de temple Hindou. Ecole Française d'Extrême-Orient, Paris

Rothermund D (2009) India. The rise of an Indian giant. Yale University Press, New Haven

Sharma R (1999) Power elite in Indian society: study of the Shekhawati region in Rajasthan. Rawat Publications, Jaipur

Sharma M (2008) Forts in Rajasthan and recent tourism inclination. S Asian J Tourism Heritage 1 (1)
Sharma S (2012) Socio-structural impact on tourism. Impact on Jaisalmer. PhD dissertation, Department of Tourism and Hotel Management, Kurukshetra University
Stern H (1991) L'Inde des familles. Enquête ethnographique. Le Rajasthan, des royaumes à l'Etat. L'Harmattan, coll. Recherches Asiatiques, Paris
Taft F (2003) Heritage hotels in Rajasthan. In: Singh S, Joshi V (eds). Rawat Publications, Institute of Rajasthan Studies, New Delhi-Jaipur
Tillotson G, Rupert H (1987) The Rajput palaces: the development of an architectural style, 1450–1750. Yale University Press, New Haven
Timberg ThA (1978) The Marwaris, from traders to industrialists. Vikas, New Delhi

Part IV
Marginalization and Spatial Reversal

Insular Oases in Globalisation: The *Ribeiras* of the Cape Verde Archipelago, Fragmented and Fragile Areas on the Way to Marginalisation

Frédéric Alexandre

Abstract Comparing oases to islands in the desert may seem a rather simple cliché. Lacoste (in *Encyclopaedia Universalis* 2014) recalled the two characteristics of oases that justify such a comparison: on the one hand, "the violent contrast between the oasis, its water and its abundant vegetation and the arid or semi-arid areas that surround it" and, on the other hand, "in desert lands […] that were travelled across for centuries, the oases were stopping-places that were vital to locate and even to control". This second feature is often lacking in oases situated on islands or archipelagos, which thus appear doubly insular and somewhat reduced in their function. This is particularly true of Cape Verde, whose "islandity", to use the term coined by Bonnemaison in L'Espace géographique 19–20(2):119–125, (1990), was strong for a long time, notably in the twentieth century during the long period of decline imposed by Salazar's Portugal. The Cape Verde islands then withdrew into their own microcosms and seemed to turn their back on their coasts. In the last decades of Portuguese colonisation and, for different reasons, in the years following independence, the Cape Verdean archipelago experienced a form of "superinsularity" (Taglioni in Annales de Géographie 652:664–687, 2006), which ended the conversion to liberalism of the Cape Verde economy and, as a result, its inclusion in the global economy. Only in the last twenty years has the Republic of Cape Verde become part of globalisation, due to the contribution of emigration and the development of services, particularly in transport and tourism, which has led to a spatial reorganisation in which the agricultural *ribeiras* have become marginalised. At the same time, beach resorts have proliferated on the islands of Boa Vista and neighbouring Sal, like "neo-oases" created *ex nihilo*.

Keywords Insularity · Oases · Cape Verde · Irrigated agriculture · Rural development · Tourist resorts

F. Alexandre (✉)
EA 7338 Pléiade - UFR LLSHS, Université Paris 13-Nord, 99 Avenue Jean-Baptiste Clément, Villetaneuse 93430, France
e-mail: alexandre@univ-paris13.fr

E. Lavie and A. Marshall (eds.), *Oases and Globalisation*,
Springer Geography, DOI 10.1007/978-3-319-50749-1_11

1 Introduction

Comparing oases to islands in the desert may seem a rather simple cliché. The origin of the metaphor is attributed to Herodotus in relation to Egypt while Ritter, in his *Géographie générale comparée* (1836, for the French translation), pointed out that the Arabs called several oasian places "island" (*Al Djazaïr*). Lacoste (in *Encyclopaedia Universalis* 2014) recalled the two characteristics of oases that justify such a comparison: on the one hand, "the violent contrast between the oasis, its water and its abundant vegetation and the arid or semi-arid areas that surround it" and, on the other hand, "in desert lands [...] that were travelled across for centuries, the oases were stopping-places that were vital to locate if not to control". This second feature is often lacking in oases situated on islands or archipelagos, which thus appear doubly insular and somewhat reduced in their function.

This is particularly true of Cape Verde, whose "islandity",[1] to use the term coined by Bonnemaison (1990), was strong for a long time, notably in the twentieth century during the long period of decline imposed by Salazar's Portugal. The Cape Verde islands thus withdrew into their own microcosms and seemed to turn their back on their coasts. In the last decades of Portuguese colonisation and, for different reasons, in the years following independence, the Cape Verdean archipelago experienced a form of "superinsularity" (Taglioni 2006), which prevented the conversion to liberalism of the Cape Verde economy and, as a result, its inclusion in the global economy.

In these islands, where the aridity increases from the wetter mountainous interiors towards the coasts, the Cape Verdean oases, constituted by the downstream part of the valleys or *ribeiras*, are very small fragmented areas. Nevertheless, a more productive agriculture is possible here, less dependent on the vagaries of the climate. In this chapter, two of these islands are taken as examples: Santiago, the most populated, and Boa Vista, the most arid. These islands were studied within the framework of the *Bilan prospectif 2004–2008*, the research section of the PRCM.[2] They were two of the study sites of the theme "*Dynamiques des paysages agricoles et forestiers*" (Dynamics of Agricultural and Forest Landscapes) (Alexandre and Saïd 2008). Fieldwork carried out in June 2007 in the context of this programme identified the particular spatial dynamics of the oasian agroecosystems on these two islands and the specific economic choices made since the liberalisation of the

[1] "Insularity is isolation. Islandity is the break with the rest of the world and in a space outside space, a place outside time, a bare place, an absolute place. There are degrees of islandity, but an island is all the more so when the break is strong or experienced as such" (Bonnemaison 1990).

[2] Regional Partnership for the Conservation of the Sea and Coastal Zone of West Africa: Mauritania, Senegal, Gambia, Guinea Bissau, Republic of Guinea and Republic of Cape Verde.

country: the abandonment of agriculture and the installation of large tourist resorts in the most arid islands of the archipelago.

Since 2007, monitoring by satellite imagery and statistics has confirmed these spatial dynamics, which have resulted in the marginalisation of agricultural *ribeiras*, despite the efforts made in terms of the integrated management of the water resource and the construction of dams in order to safeguard small irrigation schemes (da Rocha Nascimento 2013; Baptista et al. 2015). At the same time, based on an economic growth that contrasts with the rest of West Africa, Cape Verde has renewed its commitment to the development approach followed in the last twenty years: "the idea is to build a competitive economy, highly diversified and viable, through the development of seven principal economic niches" (African Development Bank 2014). The first of these niches consists of "promoting tourism with high added value". Thus, tourist enclaves are being established outside the *ribeiras*, functioning like neo-oases on the most arid islands (Sal, then Boa Vista).

2 The *Ribeiras*: A Tiny Fragmented Oasian System in a Sahelian Archipelago with a Poor Rural Economy

The Cape Verde islands, about 500 km off the coast of Dakar, are often considered a "maritime Sahel" (Marchal 1994; Lesourd 2004), fully justified by their geographic proximity to the Sahara, especially on the days when the Harmattan carries its heat and dust as far as the archipelago. Made up of a dozen volcanic islands, Cape Verde, due to its more southern latitude, has a climate characterised by a seasonal rhythm that differs from that of the other Macaronesian archipelagos. Like on the mainland, this rhythm is linked to the fluctuation of the African monsoon, with a rainy season between July and October when the intertropical convergence rises in latitude. There is strong interannual irregularity and the last fifty years have been marked by a serious decline in the winter rainfall, despite a slight recovery since the mid-1990s. The aridity is modulated between the islands, partly due to the classic contrast between the windward islands (*Ilhas Barlavento*), more exposed to the north-east trade winds, and the leeward islands (*Ilhas Sotavento*), and partly by the presence or absence of high enough mountains so that, by orographic lift, the rainfall is a little more abundant. Within the archipelago, this second characteristic distinguishes the agricultural and densely populated islands (Lesourd 1994; Location map: Fig. 1) from sparsely populated islands dominated by extensive pastureland.

Despite the maritime position of the islands, their rainfall, outside the mountainous areas, is light and irregular depending on the year, especially on those situated at the junction of the windward and leeward islands (Fig. 1). At Praia, on the island of Santiago, the mean annual precipitation is 210 mm, while it is only 80 mm at Espargos on the island of Sal (climate normals established for the end of

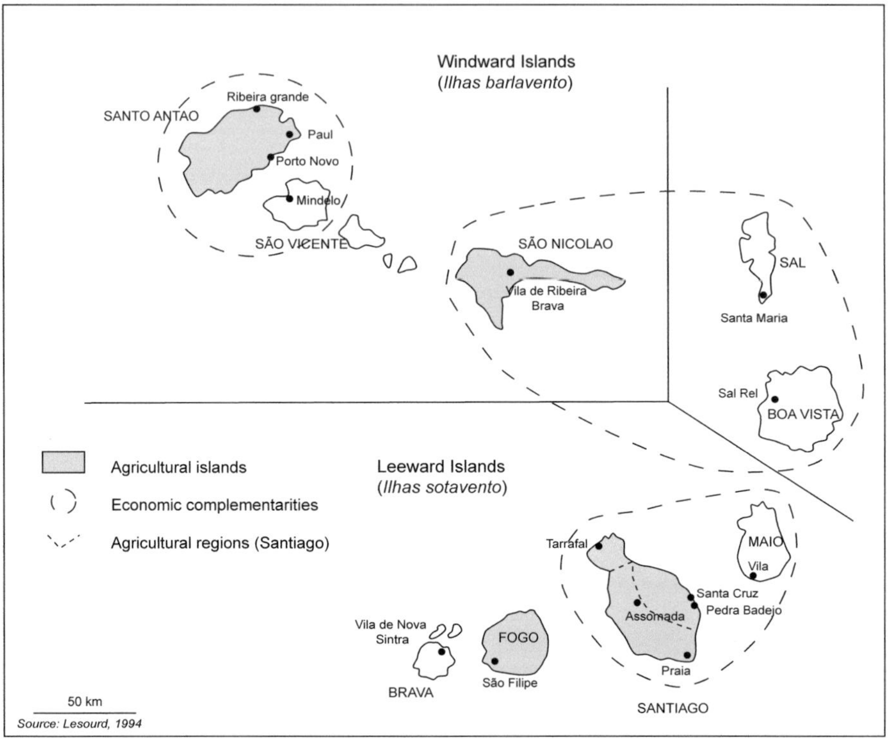

Fig. 1 Spatial organisation in the Cape Verde archipelago (modified from Lesourd 1994)

the twentieth century, ROSELT-OSS 2008). Moreover, in the second half of the twentieth century, and in the whole of the Sahel-Sudan zone of West Africa, there were two strongly contrasting periods of rainfall: from 1950 to 1970, rainfall was abundant while from 1970 to 1995, there was a succession of very dry years. For Praia, de Brum Ferreira (1987) reported a mean of 343 mm for the decade 1949–1958 and only 116 mm for the decade 1968–1977.

In this archipelago abandoned by Portugal and thus marked by its superinsularity (Taglioni 2006), an "*économie de l'indigence*" (economy of poverty) was established, to use the term of the historian Pélissier (in *Encyclopaedia Universalis* 2014). This lasted until after independence, obtained belatedly in 1975, in a difficult environmental context marked by the climate crisis (de Brum Ferreira 1987). Even today, this rural economy still concerns a large part of the population (almost 40%; Table 1), although agriculture contributes less than 10% to the GDP (African Development Bank 2014). Table 1 shows, however, the net exodus of the rural population during the last 20 years, notably to Santiago where the growth of the urban population—especially in the capital, Praia—is spectacular. These Figures are an indication of the marginalisation of rural areas in a Cape Verdean economy open to globalisation.

Table 1 Evolution of the population of Cape Verde according to 10-yearly censuses

Island	Population 1990 (% urban pop./% rural pop.)	Population 2000 (% urban pop./% rural pop.)	Population 2010 (% urban pop./% rural pop.)
Santiago (991 km^2)	175,691 (42.1%/57.9%)	236,627 (51.6%/48.9%)	274,044 (66.7%/33.3%)
Boa Vista (620 km^2)	3452 (44.1%/55.9%)	4209 (48.1%/51.9%)	9162 (59.0%/41.0%)
CAPE VERDE (4033 km^2)	341,491 (44.1%/55.9%)	434,625 (53.4%/46.6%)	491,875 (61.8%/38.2%)

Source National Institute of Statistics, Republic of Cape Verde, 2015

The "barren nature" of the Cape Verdean rural areas also explains this lack of interest after the first years following independence during which the State and the rural communities tried to rise to the challenge (Lesourd 1992). The cultivable land represents less than 10% of the territory, and the extensive pastureland is estimated at 20% while the land unsuitable for any exploitation occupies more than 50% of this same territory (*Programme d'action national de lutte contre la désertification*, National Action Programme against Desertification, 2000). The *ribeiras* make up only a small part—albeit essential—of the cultivable land.

On each island, the aridity, increasing towards the coasts (Figs. 2 and 3), concentrates the rural population on the internal plateaus and mountainsides, despite the unpredictable nature of the rain-fed agriculture (*sequeiro*) they practise there. Near the coasts, in contrast to the low-level plateaus (*achadas*) with very thin soils, often reduced to regosols, the *ribeiras*—valley bottoms of short island rivers—are small oasian areas where irrigation could be locally developed, usually in a rudimentary way. Figures 2 and 3 highlight the limited space occupied by the zones of irrigated crops in the *ribeiras* on the islands of Santiago and Boa Vista.

To ensure greater regularity in the supply of water to the *ribeiras*, hydraulic works were carried out within the framework of the water policy actively pursued by the Cape Verdean State in the 1980s and 1990s (Lesourd 1992; da Rocha Nascimento 2013). Lesourd (1992) recalled the stakes of this rural development effort:

> [The] potential is certainly modest, but the productivity of a hectare of *regadio* (irrigable plots) is 20 to 25 times higher than that of a hectare of *sequeiro* (dry farmed plots). The adventure is thus worth a try.

However, once Cape Verde had chosen to open up its economy to globalisation, the State investment was reduced in this sector of activity, which remains largely dominated by food production, generating few commercialised products. Thus, in the last twenty years, the construction of hydraulic works[3] has relied mostly on international cooperation.

[3]See the site of the Ministry of Rural Development: http://www.mdr.gov.cv/index.php/arquivonoticias/13-projectos/.

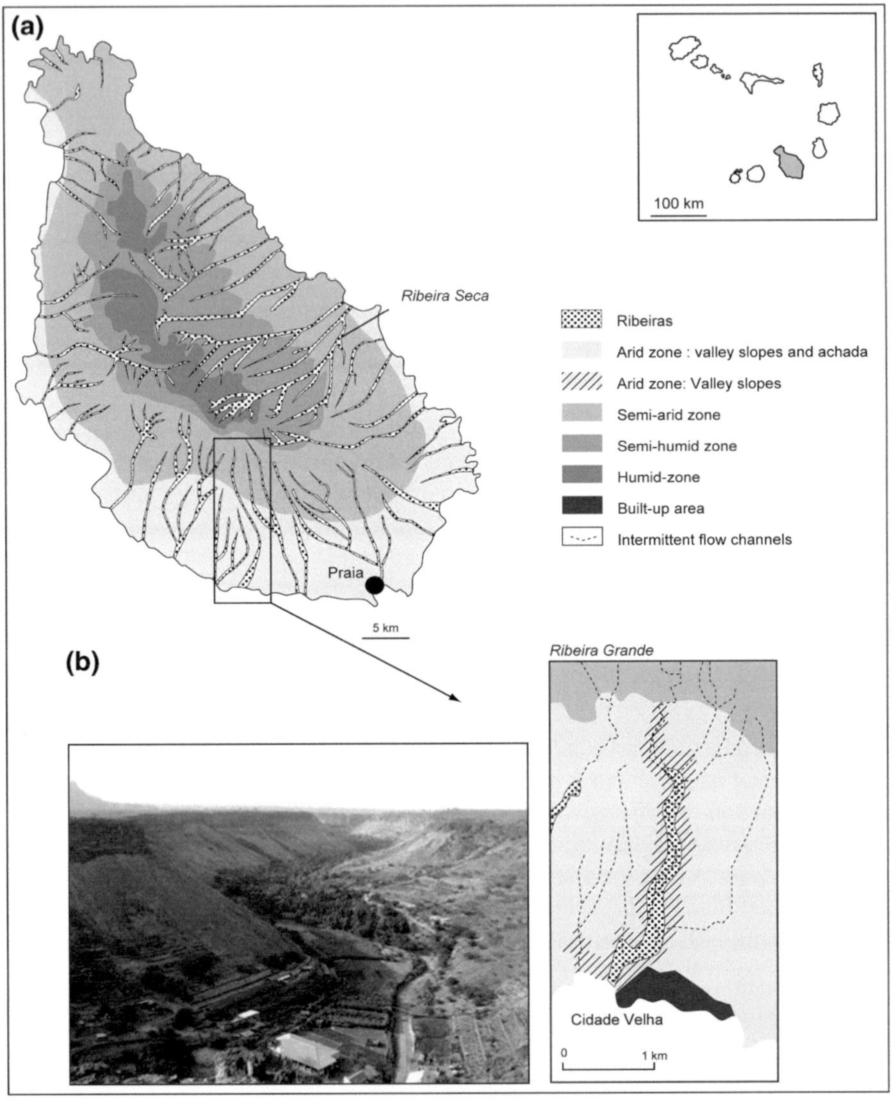

Fig. 2 Agroecological zoning on the island of Santiago (*sources* de Brum Ferreira 1987; da Silva and Granvaux Barbosa 1958; Castanheira Diniz and Cardoso de Matos 1986a) and zoom to The Ribeira Grande (south coast of Santiago): **a** agroecological map (*source* Castanheira Diniz and Cardoso de Matos 1986a); **b** the irrigated area of the Ribeira Grande seen from the citadel of Cidade Velha (*Photograph* Alexandre 2007)

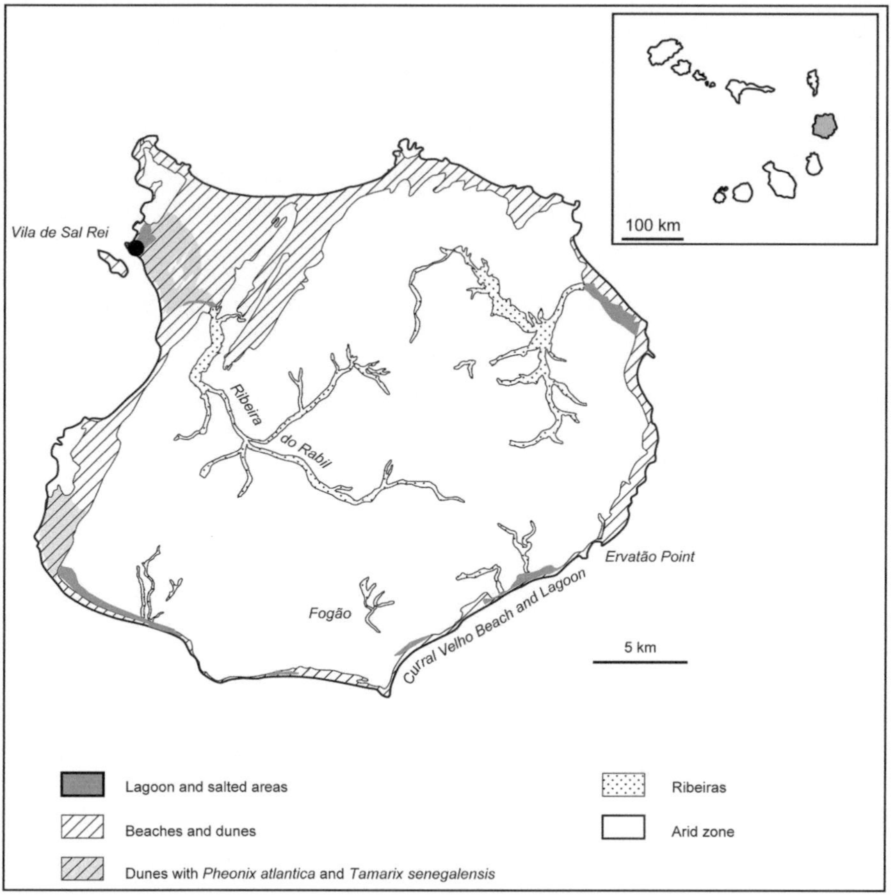

Fig. 3 Agroecological zoning on the island of Boa Vista (*source* Castanheira Diniz and Cardoso de Matos 1986b)

3 Diversity of the Oasian Archipelago of the *Ribeiras*

At the south end of the island of Santiago, the Ribeira Grande (Fig. 2) is a typical example of these fragmented areas, a sort of oasian archipelago within each Cape Verde island. The oasis is situated in the downstream part of the valley of a small river flowing from the Serra do Pico dá Antonia, which cuts deeply into an *achada*, a low basalt plateau in the arid agroecological zone (Castanheira Diniz and Cardoso de Matos 1986a, b). This low sublittoral plateau is too dry for agriculture and often appears totally bare, even though it has been largely reforested with *Prosopis juliflora*. At the estuary, obstructed by a shingle bar that prevents sea water from entering the valley, the small town of Cidade Velha is situated. It was the first capital of Santiago but is now very marginalised, located at around 10 km from the current national capital, Praia, with its 130,000 inhabitants. Observed from the ruins of the citadel of Cidade Velha, the Ribeira Grande is an irrigated, intensely cultivated area. From this viewpoint, the sequence of downstream (sugar cane, market gardening) and upstream (tropical fruit trees) crops can be seen. In general, the crops developed in the *regadios* of the *ribeiras* are based on the association of sweet potatoes, manioc and bananas (Lesourd 1992). Since 2012, the Salineiro dam has safeguarded the water supply to this irrigated area of 66 ha.

The *ribeiras* vary depending on the degree of sophistication of their irrigation system, whether they are linked to a town, and especially their size. As well as the Ribeira Grande, Santiago has several large *ribeiras*, like the Ribeira Seca on the east coast of the island, which benefited from the construction of the Poilão dam in 2006 with aid from China, thus doubling the irrigated area in the downstream part of the *ribeira* to 100 ha. Moreover, the Ribeira Seca is subject to careful monitoring: the pilot observatory of the Ribeira Seca belongs to the ROSELT-OSS (*Réseau d'Observatoires de Surveillance Écologique à Long Terme—Observatoire du Sahara and du Sahel*) programme (Network of Long-term Ecological Monitoring Observatories—Sahara and Sahel Observatory), in which the INIDA (*Instituto Nacional de Investigação e Desenvolvimento Agrário*-National Institute for Agricultural Research and Development) is very involved (Silva 2003; ROSELT-OSS 2008). The results obtained in terms of the Integrated Water Resources Management (IWRM) (da Rocha Nascimento 2013) and the strategy of Soil and Water Conservation (SWC) have given rise to hopes of a spatial generalisation of methods to other catchments in the country (Baptista et al. 2015). However, the Ribeira Seca remains an isolated example due to a lack of resources to implement such methods, and also because the *ribeiras* are generally too small and too isolated to lend themselves to such generalisation (see, in the third part, the case of the Ribeira São João).

The very incomplete nature of the Cape Verdean oases is even more obvious on the island of Boa Vista, with its much more arid climate. Boa Vista is essentially occupied by a low plateau roamed by herds of goats that, by their extensive grazing, have accentuated the desert appearance of a landscape almost devoid of vegetation where sand seas (ergs) alternate with wind-eroded plains (regs). The biogeographer Chevalier (1935) gave this description:

> the islands of salt pans are, by their appearance more than by their climate, almost Saharan lands. The wind blows continuously; the soil is bare almost everywhere from March to October. Here and there, small oases colonised mostly by Phoenix and Tamarix provide a little shade, but there is insufficient water for irrigation.

One of the two principal oases of Boa Vista is located next to its "capital", Vila de Sal Rei. It is a palm grove of the Cape Verde date palm (*Phoenix atlantica*) (Henderson et al. 2003), situated in the corridors between the dunes in the north-west of the island and associated with the long Ribeira of Rabil (Fig. 4). Its upstream section contains a few poor pastures whose growth is inhibited by the intermittent presence of water in the alluvium.

Fig. 4 North-west of the island of Boa Vista: Vila de Sal Rei and the Rabil Ribeira. Sand encroachment of the agricultural *ribeira* and proliferation of coastal tourist neo-oases (*source* Google Earth 2015; *Photographs* Alexandre 2007)

There are also many micro-oases, sometimes reduced to a garden associated with a tiny farm: on Boa Vista, they are found on either side of Ervatão point, on the east coast of the island, or in endorheic basins, like at Fogão in the centre of the island (Alexandre and Saïd 2008) (Fig. 3).

At Fogão, the run-off from the low surrounding mountains supplies an alluvial groundwater that men have exploited using pumps, previously wind-driven, today petrol-driven. Planted with Cape Verde date palms (*Phoenix atlantica*) and tamarinds (*Tamarindus indica*) strongly deformed by the wind, the oasis is reduced to a few irrigated plots where maize, marrows, carrots, cabbages, etc., are grown. The cultivated plots are protected by high walls, topped with thorns; there are a great many goats. The agricultural workers encountered said that the owner also raises cattle and pigs. Nevertheless, this farming is fragile: according to them, it is barely self-sufficient in dry years while in wet years, some of the production can be put on the market.

4 From the Legacy of the Portuguese Colonisation to an Archipelago Open to the Market Economy: Marginalisation of the Agricultural *Ribeiras*

Although the climatic constraints that affect the archipelago and the small size of the agricultural territory both play an undeniable role, the situation in the Cape Verde countryside is also due to the legacy of the Portuguese colonisation, witnessed by the minor and belated investment in improving the irrigation systems. The Portuguese colonisation favoured the seizing of land by the big landowners, especially in those sectors where there was water. This land issue, that land reform in the years following independence tried to address, also contributes to the special nature of Cap Verdean oases.

Auguste Chevalier (*op. cit.*) provided a careful description of the archipelago in 1935 when there was a great deal of land grabbing by big landowners. Representing well the colonial science, Auguste Chevalier regrets that the noble Portuguese nation "is slow, on these islands, to redirect colonisation toward new pathways, conserving nature more, and ceasing to ransack the soil and the vegetation of tropical lands". He also recalls the extent of the clearing of the archipelago, which completely transformed the area into cultivated fields or grazing for livestock, without meeting the needs of a population regularly decimated by famine, closely linked to the recurrent droughts: 1902–1904, 1920–1923, 1941–1942, 1946–1948 and 1959. Pélissier (in *Encyclopaedia Universalis* 2014) points out the indifference of the Salazar regime, or even the advantage it found in such a situation:

> by inertia or calculation, the Portugal of 1920 to 1950 did nothing to alleviate the local problems, neither drilling wells nor reforesting, so that the fear of famine no longer channelled emigration towards the United States, but towards Portuguese Guinea, São Tomé and Angola where willing local labour was lacking. By reducing public investment to a minimum […], the Portugal of Salazar found in this archipelago a central support and a reservoir of labour for the colonies.

After independence, in response to this inaction, the political authorities were anxious to manage the meagre resources of the territory better, by a voluntarist policy of rural development based on searching for and exploiting water resources and on strengthening the fight against desertification and erosion (Lesourd 1989). Implemented in a difficult climatic context after the succession of dry years in the 1970s and 1980s, this policy relied on community projects so that a population deprived of work and the means to survive could earn a small extra income. These works—which at the time also benefited from international aid, particularly from the FAO—included the maintenance of roads, the building of small dams to develop the irrigated areas of the *ribeiras* (Lesourd 1992) and reforestation, which was huge in the most arid agroecological zone (Fig. 1), on the *achadas*, plateaus that border the ocean (Castanheira Diniz and Cardoso de Matos 1986a, b; Spaak 1990). Thus, "from 1976 to 1985, almost 11 million trees were planted over a total area of 24,191 ha. From 1986 to 1990, reforestation continued at a rapid pace" (Courel and Chamard 1992). The plantations used notably a Mimosaceae of American origin, *Prosopis juliflora*, which has become a pantropical species, widespread from Cape Verde and Senegal to Cameroon and as far as Somalia. It is a small thorny tree (6–8 m), shrubby or with a short bole (Arbonnier 2002). When the terrain is sloping, these plantations have some means to limit erosion, for example dry stone walls to retain a little soil.

Thirty years later, the mobilisation of rural development seems a long way off and the works that resulted are now being questioned. For example, in the village of São João Batista (Ribeira São João, Santiago), at the bottom of a small *ribeira* deeply cut into the basalt *achada*, the plantations of *Prosopis juliflora* are seen by the inhabitants as reinforcing their isolation. Moreover, their value is contested as the villagers have only limited access to the acceptable extra fodder they could provide (Andrieu and Alexandre 2010). The young people present could not name the tree that is the key element of their immediate environment, even under its more common name of *Acacia americana*. They pointed out the lack of jobs and future prospects in a sector where the agricultural production is considered insufficient to meet the food needs of a still significant population. In this *ribeira*, irrigation is supplied by drilling and managed by the State: each farmer can have a part of this water, provided he pays for the fuel to operate the pump. Only one significant source of extra income was mentioned by the villagers: the sale of building materials, gravel and stones, removed, not without consequences, from the shingle bar that closes the *ribeira* downstream. This removal of sand and gravel is a major environmental problem for the *ribeiras* as it weakens the sand and shingle bars allowing salt water to flow into the low valleys.

Twenty years ago, Marchal (1994) highlighted the impossible equation between a rapidly growing population, still 70% rural, and an agriculture that hardly covered more than 40% of the country's food needs in the best years and less than 10% during the most severe droughts. Today, although the strategic document published by the African Development Bank in 2014 continues to affirm that "agriculture remains a crucial sector for the reduction of poverty, the promotion of green growth and climate resilience", it remains essentially food production and represents scarcely 10% of the GDP. This has led to the marginalisation of rural areas, particularly marked in the *ribeiras*, and although the proportion of the rural population has fallen significantly, it is still almost as large in absolute numbers (Table 1). The rural exodus will only begin to have an effect from the beginning of the third millennium, especially on the overpopulated high plateaus, devoted to rain-fed agriculture (cf. Figures from the *conselho* of Santa Catarina), much less in the *ribeiras*.

In addition, the voluntarist policy of rural development relied on a land reform (applied since 1982) already described as having "mixed results" in 1994 by Michel Lesourd. It has certainly produced some outcomes, such as eliminating the risk of famine, but it is no longer a priority in the political and economic choices of the country, which has led to the spatial marginalisation of agricultural *ribeiras*, particularly the small, poorly intensified ones. Signs of this marginalisation are the significant environmental problems that have arisen (choking of palm groves with sand, salination of parts of *ribeiras* close to the coast, soil exhaustion, etc.).

It is also striking to see that the pilot *ribeiras* in terms of rural development and environmental monitoring, like the Ribeira Seca (Santiago) mentioned above, are not spared these problems. Diachronic analysis of soil use shows significant changes in the intensification of the agroecosystem during the last fifty years (Silva 2003). The plantations of *Prosopis juliflora* have reached the *achadas,* and the dry agroforestry has spread over the slopes. The irrigated zones have spread in the downstream part, due to the construction of the Poilão dam (da Rocha Nascimento 2013). However, during our visit in 2007, the environment of the *ribeira* appeared degraded and it will be a struggle to achieve the objectives there. At the river mouth (Foz de Ribeira Seca), the sandbar that closes the *ribeira*, on which the Senegal Tamarisk (*Tamarix senegalensis*) has become almost the only species to the detriment of the original vegetation, has been breached, due to the removal of too much sand for construction in the nearby town of Pedra Badejo. As a result, salt water has entered the lower part of the *ribeira*. In addition, the drilling carried out has allowed the gradual intrusion of the saltwater table towards the interior of the valley. Furthermore, freshwater could become even less abundant at the river mouth as it is retained upstream by the Poilão dam (Alexandre and Saïd 2008).

The degraded state of the environment is also visible in the north-west part of Boa Vista. As mentioned above, the corridors between the dunes of the largest sand sea on the island, where the freshwater table is close, are occupied by a palm grove of Cape Verde date palms (*Phoenix atlantica*). However, the sand of the dunes can still move and even the corridors between them contain small mobile dunes that choke the palm trees, as already noted by Auguste Chevalier in 1935. *Phoenix atlantica* is a demanding species, requiring the constant presence of freshwater close to the surface. Thus, plantations of *Prosopis juliflora* were established to try to stabilise the sand. This shrub has become invasive, dispersed by the livestock that consume its pods, and now strongly competes with date palms for the water resource. Further south, the Rabil Ribeira is also affected by the progress of the dunes (Fig. 4) while the downstream part is strongly salinated due to poor maintenance of the dams that confine the sea water to a short stretch at the mouth of the *ribeira*.

In short, in the Cape Verdean context of insertion of the economy into globalisation, the double insularity of the *ribeiras* finds itself strengthened. The spatial consequences of the economic liberalisation vary depending on the islands. In Santiago, it mainly shows through urban growth, especially at Praia. In Boa Vista, tourism obviously expands. We are thus led to show that a new form of oases emerges around touristic resorts, out of the *ribeiras* case.

Indeed, one of the major results of the spectacular conversion of the Republic of Cape Verde to a liberal economic policy, largely open to globalisation, has been the development of tourism on the archipelago. It began with seaside tourism for an international clientele on the island of Sal, well provided with sandy beaches as well as having the archipelago's largest airport at Espargos. This tourism has diversified, reflected by "notably, varying numbers of visitors, differing levels of integration of tourist sites with living spaces, and economic, social and environmental impacts that differ according to the island" (Kokel 2009). Nevertheless, the quantitative (Table 2) and economic assessment of tourism development relies essentially on the proliferation of beach resorts on the most arid islands of the archipelago. Although they do not correspond to the usual criteria for oases (see the introduction), these tourist enclaves can be described as neo-oases or pseudo-oases. In fact, their only typical oasian characteristic is their landscape that contrasts strongly with the desert environment.

After Sal, these tourist resorts have multiplied in recent years on the west coast of Boa Vista, in stark contrast with the *ribeiras*, small agricultural oases, inheritors of "the economy of poverty" (Fig. 4). Moreover, during the 2000s, Boa Vista overtook Sal in terms of accommodation capacity and nights spent on the island (Table 2). Between them, Sal and Boa Vista represented almost 75% of tourist visits to the archipelago in 2011 (35.4% for Sal and 38.9% for Boa Vista).

Foreign investors were called upon to build these hotel complexes, but the improved infrastructures needed for tourism development—motorways, roads, desalination units providing drinking water—were only made possible by diverting public investment towards these two islands, where only a tiny part of the Cape Verdean population, and an even smaller proportion of the rural population, lives. In the 2010 census, Sal had 25,000 inhabitants and Boa Vista less than 10,000, while the total population of the archipelago was 491,000 inhabitants, 188,000 of them rural (Table 1).

In 2007, the first tourist resort of more than 300 rooms and developed by Italian investors was noted stretching along Chave beach (Fig. 4). It was built on the *achada* south of the Rabil Ribeira, close to the airport of Boa Vista, and was recently renovated to meet international standards. Since then, a second has opened a few kilometres further south and a rapid inventory reveals six more already built or under construction in 2015.

Foreign investment projects are increasing on the west coast and now on the south coast of the island. The public authorities have just tarred the road between the airport of Rabil and a new tourist enclave in the south of Boa Vista. The name of this new resort—"*Riu Tuareg*"—plays on the image of the Saharan oasis, which is reached after crossing the austere plateaus of Boa Vista with their landscapes

Table 2 Development of the tourism sector in Cape Verde from 2000 to 2011 and concentration of the provision on the islands of Sal and Boa Vista

	Total Cape Verde		Santiago		Sal		Boa Vista	
	Entries	Nights	Entries	Nights	Entries	Nights	Entries	Nights
2000	145,076	684,733	30,514	98,515	75,016	424,276	9402	63,161
2001	162,095	805,924	28,196	92,564	93,496	548,901	10,168	67,533
2002	152,032	693,658	24,514	84,195	93,783	478,385	9023	58,541
2003	178,379	902,658	28,421	92,037	116,319	685,511	7918	48,195
2004	184,738	865,125	24,756	80,830	129,608	685,198	3849	22,129
2005	233,548	935,505	33,501	107,369	162,625	709,982	4582	24,306
2006	280,582	1,368,018	55,648	139,501	167,222	945,421	20,698	171,866
2007	312,880	1,432,746	60,786	133,098	192,038	1,101,642	15,533	90,796
2008	333,354	1,827,196	67,107	137,063	190,137	1,347,076	33,135	238,720
2009	330,319	2,021,597	53,752	120,192	148,005	1,073,300	82,476	705,188
2010	381,831	2,342,282	52,110	114,804	154,115	1,104,004	125,575	1,000,271
2011	475,294	2,827,562	56,693	130,632	168,322	1,214,066	184,878	1,334,108

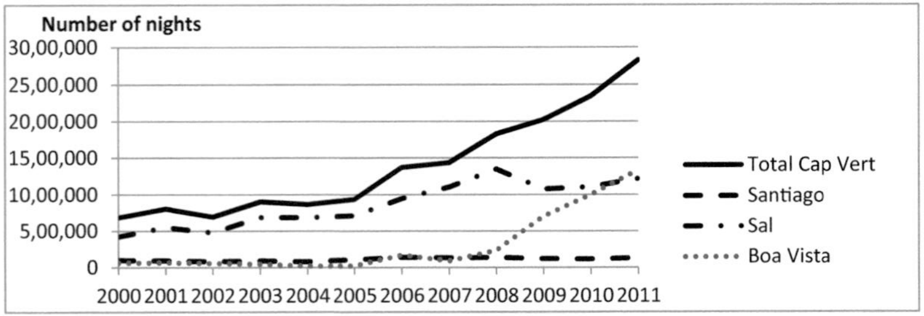

Source National Institute of Statistics, Republic of Cape Verde 2014

alternating between *regs* and *hamadas*. Such a construction inevitably raises some questions, for example about the water supply and the evolution of the coastline. In fact, the coastal drift is strong near this tourist resort.

Further away, another well-advanced project concerns the beach of Santa Monica, affecting the south-east coast of the island classified as a nature reserve because of the fragility of its sandbar and lagoon. Moreover, it is a reproduction zone for sea turtles, especially in the lagoon of Curral Velho. In the current tourism development plan, the eastern coasts of Boa Vista are preserved and only eco-tourism can be practiced there.

5 Conclusion

The *ribeiras* of Cape Verde, small irrigated cultivated enclaves on islands where the aridity makes rain-fed agriculture very unpredictable, constitute an original form of oasis. This originality is due to their fragmentation and isolation, reinforced by their insular setting. Retaillé (2003) highlights the sophistication of the technical and social systems that have generally underpinned the setting up of oases, as well as the mastery of the arid area that their success implies. In this regard, the Cape Verdean *ribeiras* seem rather poor oases. They are not agroecosystems with a long rural history, which have established, sometimes since antiquity, clever and complex irrigation networks.

However, in the years following independence, the *ribeiras* benefited from the efforts made by the new authorities in favour of a rural development that then seemed essential for a young republic wanting to be self-sufficient: land reform, development of irrigation and the Integrated Water Resources Management (IWRM), the fight against soil erosion on the mountainsides, etc. The new political and economic choices made in the 1990s, in favour of liberalism and opening up to a market economy, radically changed things. The inclusion of the Cape Verde Republic in globalisation, founded on the contribution of emigration and the development of services, particularly in transport and tourism, has led to a spatial reorganisation in which all rural areas, including the agricultural *ribeiras*, are marginalised. The degradation of the environment in some of these *ribeiras* is a sign of this marginalisation, notably when it seems to be the result of urbanisation. For example, the removal of building materials from the sand and shingle bars closing the *ribeiras* has led to the salination of their downstream parts. Yet the resources provided by these areas remain essential in a country where the life of the rural population, which is not falling in absolute numbers and has little access to imported products, remains precarious.

At the same time, the country's inclusion in globalisation has benefited tourism, which has risen from 18% of GDP in 2006 to 25% in 2013, becoming the leading economic sector of the archipelago based on the construction of tourist beach resorts for an international clientele. The island of Boa Vista, after neighbouring Sal, has seen a growth in tourist enclaves, like neo-oases created *ex nihilo*. These implantations pose environmental problems that differ greatly from those caused by the *ribeiras*, mainly because of the high water consumption they demand in these islands whose water supply is limited.

References

Alexandre F, Saïd AR (2008). Axe 4: Dynamiques paysagères et agricoles (Rapport Bilan Prospectif PRCM - Afrique de l'Ouest - 2005-2008). Dakar, IRD, 110 p

Andrieu J, Alexandre F (2010) Dynamique des paysages forestiers sur la frange littorale de l'Afrique de l'Ouest (Sénégal, Gambie et Guinée-Bissau). In: Galochet M, Glon E (dir) Des milieux aux territoires forestiers, Mélanges en l'honneur de Jean-Jacques Dubois (actes du colloque international de Villeneuve-d'Ascq, septembre 2008). Artois Presses Université, Arras, pp 199–219

Arbonnier M (2002) Arbres, arbustes et lianes des zones sèches d'Afrique de l'Ouest, 2e éd. CIRAD/MNHN, Paris, 573 p

Banque Africaine de Développement (2014) Cabo Verde: document de stratégie pays 2014-2018. 24 p. + annexes

Baptista I, Fleskens L, Ritsema et al (2015) Soil and water conservation strategies in Cape Verde and their impacts on livelihoods. an overview from the Ribeira Seca Watershed. Land 4:22–44. doi:10.3390/land4010022

Bonnemaison J (1990) Vivre dans l'île. Une approche de l'îléité océanienne. L'Espace géographique, Tome 19–20, n° 2:119–125

Castanheira Diniz A, Cardoso de Matos G (1986a) Carta de zonagem agro-ecologica e da vegetação de Cabo Verde. I- Ilha de Santiago. Garcia de Orta, Série de Botânica 8, n° 1–2 (tirage à part):39–82 + carte h.-t

Castanheira Diniz A, Cardoso de Matos G (1986b) Carta de zonagem agro-ecologica e da vegetação de Cabo Verde. IV- Ilha de Boavista. Garcia de Orta, Série de Botânica 10, n° 1–2 (tirage à part):49–72 + carte h.-t

Chevalier A (1935). Les Iles du Cap-Vert. Géographie, biogéographie, agriculture. Flore de l'Archipel. Revue de botanique appliquée et d'agriculture coloniale 15, n° 170–171:733–1090

Courel M-F, Chamard P (1992) Reforestation au Cap-Vert. Afrique contemporaine n° 161 (1er trimestre 1992):238–247

da Rocha Nascimento J (2013). Le bassin de Ribeira Seca de Santiago, Cap-Vert: une gestion basée sur l'approche GIRE. Vegueta, Anuario de la Faculdad de Geografía e Historia n° 13:95–118

da Silva Teixeira A, Granvaux Barbosa A (1958) A agricultura do Arquipélago de Cabo Verde. Memorias da Junta de Investigações do Ultramar, Lisbonne, 2a série, n° 2, 178 p

de Brum Ferreira D (1987) La crise actuelle dans l'archipel du Cap-Vert. Quelques aspects du problème dans l'île de Santiago. Finisterra, Tome XXII, n° 43:113–152

Henderson S, Gomes I, Gomes S, Baker W (2003) Phoenix in the Cape Verde Islands. Palms 47–1:5–14

Kokel N (2009) Le Cap-Vert : mise en tourisme et enjeux de développement d'un petit espace archipélagique. Thése de doctorat en géographie, Université de La Rochelle

Lacoste Y (2014). Oasis. In: Encyclopaedia Universalis, édition électronique

Lesourd M (1989) Construction nationale et insularité en milieu sahélien: la République du Cap-Vert. In: Antheaum B, Blanc-Pamard C, Chaléard J-L, Dubresson A, Lassailly-Jacob V, Marchal J-Y, Pillet-Schwartz A-M, Pourtier R, Raison J-P, Sevin O (éds) Tropiques, lieux et liens. Florilège offert à Paul Pélissier et Gilles Sautter. Editions de l'ORSTOM (Coll. Didactiques), Paris, pp 422–434

Lesourd M (1992). Maîtrise et valorisation de l'eau aux îles du Cap-Vert. Le défi d'un État et de ses paysans à une nature ingrate. In: Le Floc'h E, Grouzis M, Cornet A, Bille J-C (éds)

L'aridité, une contrainte au développement. Caractérisation, réponses biologiques, stratégies des sociétés. ORSTOM éditions (coll. Actiques), Paris, pp 520–537
Lesourd M (1994) Insularismes et développement en République du Cap-Vert. Lusotopie n ° 1:113–133
Lesourd M (2004) Cap-Vert: insularité et développement. Géopolitique Africaine n° 14 (n° spécial L'Afrique lusophone, trente ans après la Révolution des Œillets):153–164
Marchal J-Y (1994) Cap-Vert: 380 000 sahéliens en mer. In: Dubresson A, Marchal J-Y, Raison J-P (éds) Les Afriques au Sud du Sahara, tome de la Géographie Universelle sous la direction de Roger Brunet, Paris/Montpellier, Belin/Reclus, pp 117–118
Pélissier R (2014) Cap-Vert. In: Encyclopaedia Universalis, édition électronique
Retaillé D (2003) Oasis. In: Lévy J, Lussault M (éds) Dictionnaire de la Géographie et de l'Espace des Sociétés, Paris, Belin, pp 671–672
Ritter C (1836) Géographie générale comparée ou Science de la terre dans ses rapports avec la nature et l'histoire de l'Homme. Première partie: l'Afrique, traduite en français par E. Buret & E. Desor. Paris, Paulin, vol 1, 544 p
ROSELT-OSS (2008) Synthèse Afrique de l'Ouest: Flore – Végétation – Occupation des Terres. Université Abdou Moumouni, Niamey, 76 p
Silva GCC (2003) Land use system in the Ribeira Seca Basin: how to evaluate in the last half-century changes? Acta Universitatis Carolinae Environmentalica 17:59–73
Spaak J-D (1990) Boiser les îles du Cap-Vert … Pourquoi, comment, pour qui? Bois et Forêts des Tropiques 225 (3e trimestre 1990):47–54
Taglioni F (2006) Les petits espaces insulaires face à la variabilité de leur insularité et de leur statut politique. Annales de Géographie n° 652:664–687

Liwa: The Mutation of an Agricultural Oasis into a Strategic Reserve Dedicated to a Secure Water Supply for Abu Dhabi

Alain Cariou

Abstract The population of the Arabic shore of the Arabian Gulf now lives in a new urban world based on oil and post-oil production and Western models. Such is the case of Abu Dhabi, a new metropolis that now forms part of the globalised world. The spectacular growth of this capital has profoundly upset the world of the oasis. The oasis of Liwa, which formerly constituted the political and economic heart of the region, is now relegated to a peripheral role subservient to the urbanisation of the coast. This reversal of the space represents a break in the perception, the practices and the functions devolved to oasis spaces. It raises the question of the evolution of the relationships between the city and the oasis. The authorities of the United Arab Emirates have tried in vain for several decades to develop farming in the desert to provide food for the country. For Liwa, as for many oases of the Arabian Peninsula, this agricultural development policy has brought about a depletion of water resources due to the overexploitation of groundwater. In spite of the development of seawater desalination, the country faces a reduced availability of water due to the explosion in demand. Priority is no longer given to agriculture but to increasing the security of the urban supply, particularly by creating strategic reserves of water to reduce the vulnerability of the country. This is why the oasis of Liwa has been transformed into a strategic space, containing an underground artificial reservoir.

Keywords Oasis · Liwa · United Arab Emirates · Abu Dhabi · Strategic water reservoir · Water use policy · Groundwater management · Globalisation

A. Cariou (✉)
Laboratoire ENeC (Espaces, Nature et Culture, UMR 8185),
University Paris-Sorbonne, 195 Rue Saint-Jacques, 75005 Paris, France
e-mail: alain.cariou@paris-sorbonne.fr

E. Lavie and A. Marshall (eds.), *Oases and Globalisation*,
Springer Geography, DOI 10.1007/978-3-319-50749-1_12

1 Introduction

The recent emergence of Abu Dhabi as a new urban centre integrated into worldwide networks has deeply altered the hinterland oasian territories. This is the case in Liwa, whose cluster of oases used to be the economic and political centre point of the region and which is now relegated to the status of a peripheral area dedicated to the metropolisation of the coast. The priority activity of this area is no longer agriculture but the guarantee of a secure water supply to the cities, especially by using Liwa's aquifer as a strategic water reserve, thus reducing the vulnerability of the country. It now seems pertinent to study the processes responsible for this spatial shift and to measure the changes in the activities, the functions and the perception of these new oasian lands. It is widely recognised that the oasis of Liwa, like any oasis in the world, from the past to today, has always been imbedded in a vital network of relationships with remote lands and hubs. The oasis only exists as a "by-product" of a wider spatial system (Côte 2002). The current analysis shows how the role of the oasis in the spatial organisation of the country has rapidly changed since the onset of globalisation in this part of the Gulf. First, this point of view will be examined in the context of the traditional management mode of the desert area where Liwa constituted an organisational centre. Secondly, this study will focus on the processes of the change in status of the oasian space resulting from the emergence of the world metropolis, Abu Dhabi. In conclusion, the types of relationships between this "global city" and its oasian periphery will be discussed.

2 Liwa, Centre of the Traditional Organisation of Desert Lands

2.1 An Intra-desert Oasis

Located a hundred kilometres inland from the shores of the Arabo–Persian Gulf, the Liwa region is an oasian cluster in the hinterland of the Emirate of Abu Dhabi. Palm groves growing in the relatively protective interdunary depressions spread from east to west over about 70 km in a vast crescent-shaped discontinuous oasian area. This environment is part of a wide erg covering 74% of the total surface area of the Arab Emirates and is the north-eastern end of the immense sand desert of Rub al Khali (Fig. 1).

In Liwa, the accumulation of sand has gained in volume and in the south, it forms mega-barkhanes with heights reaching over 50 m, sometimes over 100 m. In this warm tropical desert, the climate is hyper-arid in type, as precipitation does not exceed 40 mm/year, whereas potential evaporation comes close to 3500 mm/year (National Atlas 1993). Rainfall displays an erratic pattern, mainly occurring in winter (30 mm) and practically never during the hotter summer season (less than 5 mm), which has an average temperature of 35 °C. With no surface run-off, life in

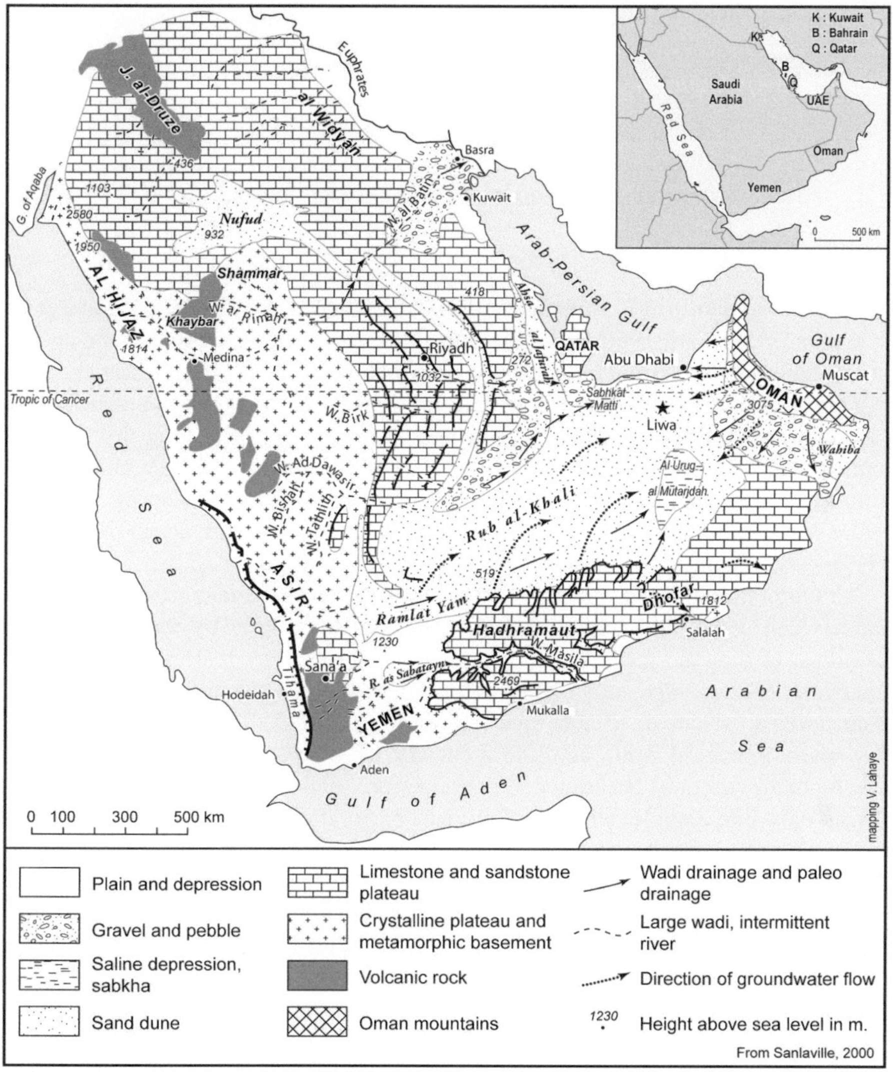

Fig. 1 Liwa, the physical environment

the oasis depends entirely on underground water layers. The presence of a thick sand cover dating from the Quaternary provided an excellent context for the development of a sandy dune aquifer with a shallow water table. The high porosity of the paleodunes (38%) thus enabled the storage of precipitation, which was about 200 ± 50 mm/year during the rainy episodes at the end of the Pleistocene, between 32,000 and 26,000 years BP, and during the Holocene between 9000 and 6000 years BP (Wood and Imes 2003; Wood et al. 2003). In the current hyper-arid climatic conditions, the turnover rate of the Quaternary aquifer is insignificant, probably less than 5%. Therefore, the presence of this mostly fossil water layer, easily accessible through wells dug a few metres deep in interdunary hollows, enabled ancient human populations to settle in these harsh desert surroundings.

2.2 Liwa, Ancient Economic and Political Heart of the Desert

In this eastern part of the Arabian Peninsula, the oasis of Liwa was the nodal point of economic life until 1966, when the onset of the oil era began to alter the ancient traditional socio-economic order. For several centuries, the oasian arc was the heart of a territorial organisation spreading from the shores of the Arabo–Persian Gulf to the fringes of the large sandy masses in the hinterland. Since at least the sixteenth century, the largest Bedouin tribe of the area, the Bani Yas, has established their summer camp in the cluster of small palm groves of Liwa, constituting an oasian centre with political control over the vast surrounding desert areas (Heard-Bey 1997).

Although the traditional palm groves of Liwa are substantial agrosystems where hamlets can settle, their presence cannot be explained only by their agricultural activities, as they are just one element of a wider spatial system based on trading, economic relationships and complementarities. Due to the physical conditions, characterised by narrow tillable areas constantly threatened by sand accumulation, the near-absence of soil, and the intensity of evapotranspiration, agricultural development can only be limited. Thus, date production, alongside fishing activities, is only one complement to a dominant pastoral economy. Dromedaries are a source of milk, meat and leather, and they are also used for riding and transport purposes. They are an indicator of civilisation as well as an identity symbol for the Bedouins. In an environment lacking resources, complementarity between ecosystems and activities was essential for the traditional economy of the Emirate Bedouins (Wilkinson 1977; Cordes and Scholz 1980). During summertime, the Bedouins used to congregate in Liwa for date harvests, while the herds would stay on the outskirts of the oases under the supervision of young shepherds. In October, the Bedouins would resume their seasonal migration to reach the sandy ergs, turned by winter rainfall into ephemeral pasturelands vital for dromedary herds (Fig. 2). Some nomadic groups would reach the Gulf coast where they would set up camp and fish. Dried and salted fish constituted a sought-after trade product, a high source of proteins, and complementary to a diet mostly based on dates. Pearl fishing from June to October, when the waters of the Gulf had warmed up, was another source of income until this activity collapsed in the 1930s due to the growth of pearl culture.

Liwa is also the cradle of a dynasty of Sheikhs whose descendants are currently ruling the Emirates of Abu Dhabi and Dubai. From the end of the eighteenth century, a part of the Bani Yas settled on the coast. The Al Bu Falah section, led by the Al Nahyan family, founded a camp around a spring, which would later become

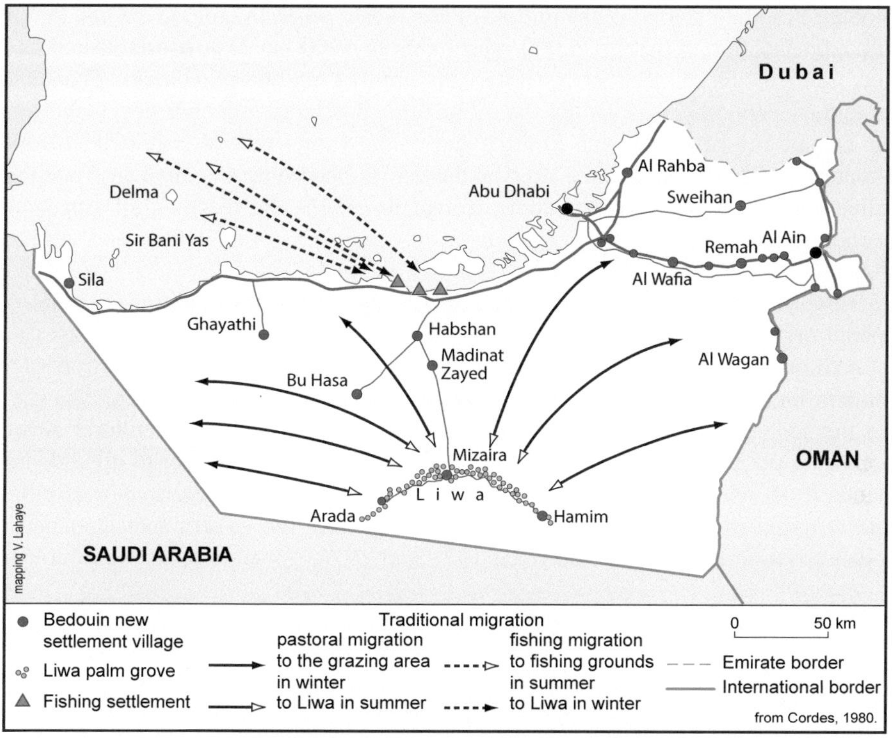

Fig. 2 Liwa oasis in the heart of a traditional economy

Abu Dhabi. Another group, the Al Bu Falasah, under the authority of the Al Maktum family, settled on the shores of a sandy bay that would become Dubai. The remarkable destiny of these Bedouin camps, which rose to the rank of a global city in the space of a few decades, has radically changed the traditional hierarchy in the organisation of space. From a central point, Liwa progressively downshifted to the status of a periphery dominated by the coastal metropolisation.

3 Reversal of Space Functionality and Marginalisation of the Oasis

3.1 Coastal Metropolisation and Sedentarisation

Despite a lack of urban traditions, the population of the United Arab Emirates is now established in city life. City dwellers, representing 80% of the total population of the Federation, are concentrated almost exclusively along the coast, leaving an almost empty desert hinterland. This coastal polarisation is the result of globalisation that started in the middle of the 1960s, when the region entered the oil era. Abu Dhabi had barely 25,000 inhabitants in 1950 but is now established as a new political and oil-producing centre with almost 1 million inhabitants. Thanks to its tertiary economy, enabled by the oil income, this metropolis has developed into the capital city of the oil-rich Emirate and of the young federal state (Cadène and Dumortier 2011). Abu Dhabi is now on the way to becoming a "global city" with an influence and an attraction spreading around the world and reaching all aspects of economic life: financial flows, movement of consumer goods, labour and tourism flows.

This spectacular urban growth has drastically altered the traditional model of spatial organisation (Fig. 3).

Although the oil revenues were massively invested in the construction of a modern metropolis integrated into the process of globalisation, they were also used for the sedentarisation of nomads and for the promotion of an agricultural development supposed to ensure an illusory secure food supply. In order to upgrade the standard of living of the Bedouin populations who lived in precarious conditions and to avoid the total depopulation of the hinterland, the government launched a sedentarisation programme. Between 1971 and 1976, 73 settlements were built in

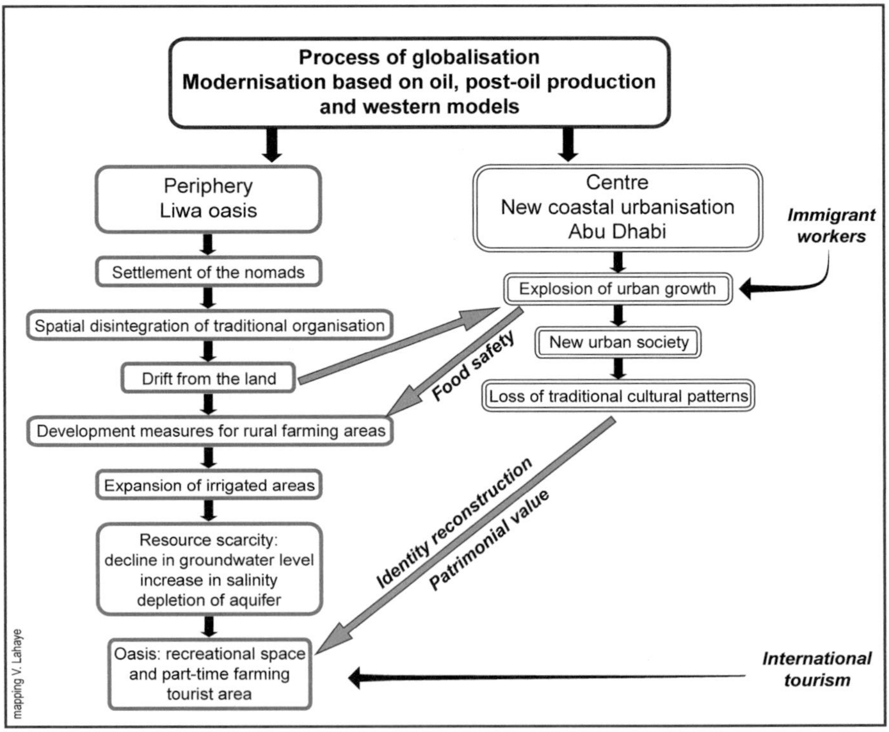

Fig. 3 The reversal of space

the oases and on the nomad campsites. Houses with running water and electricity were provided free to the local populations, who also benefited from community infrastructures: schools, health centres, shops and agricultural cooperatives dedicated to the reinforcement of livestock farming and agriculture (Cordes and Scholz 1980). In the meantime, the families were given areas of irrigated land for free. Although the transition from nomadism to sedentarisation was not continuous but a step process including semi-nomadism and transhumance, the Bedouins finally entered the urban environment, mostly giving up their sedentarisation villages in order to fit into the economic ways of large coastal cities. The attraction of material comfort offered by a modern metropolis, together with employment opportunities arranged in public administration and trade, triggered a massive rural exodus. Even if nomadic life has yielded to modern urban life, the oasian agriculture has not yet disappeared, although it has undergone deep changes.

3.2 The Dead-End of Agricultural Development

In spite of a proactive policy deployed for several decades by the authorities to promote agriculture, one cannot but notice the dead end where this sector stands today. In the original oases, as in the modern irrigated perimeters, many plots reveal how little care is given to the crops, when these are not simply abandoned. In 2011, the agriculture sector (agriculture, agroforestry and green spaces) absorbed 72% of the water consumed by the Emirate of Abu Dhabi, but contributed slightly less than 1% to the GDP. Food self-sufficiency, which used to be one of the Government's priorities, is far from being ensured, as a consequence of the abandonment of the non-profitable activity of cereal agriculture, while fruit and vegetable production only covers 15% of the demand, and the rest must be imported (EAD 2012). Still, as in all the neighbouring states of the Peninsula, the government deployed a great deal of effort in order to enhance the emergence of an agricultural periphery, able to respond to the constantly increasing urban food demand. In Liwa, the modernisation of the ancient oasian centres and the creation of new irrigated perimeters contributed to increasing the cultivated surface area from 1000 to 21,000 ha between 1987 and 2002 (ERWDA 2004). Reclaimed desert lands equipped with drilling wells by the State were given free to all the nationals (rural as well as urban) who made a request, generating multiple small-sized private family farms. Each farm, dependent on an irrigated perimeter protected by a windbreak hedge, is a square plot with an average surface area of 3 ha. Between date palms loosely planted, the soil generally bears fodder crops for dromedaries and small livestock (sheep and goats). Vegetable crops occupy less than 15% of the cultivated lands. Once pumped out from drilling wells, water is carried to the plots through cemented canals. The State has also provided service and extension centres. Every farmer has the opportunity to obtain advice from agricultural technicians, to hire equipment and to access 50% of state-subsidised inputs (fertilisers, selected seeds and phytosanitary products).

In spite of the huge financial and human means that the State has deployed over the past 40 years, the agriculture policy appears to be an economic failure, even if improvements have incontestably been made in agronomy. This is particularly the case with large capitalist horticultural units, stimulated by the high urban demand for fresh products. Throughout the year, they produce strawberries, lettuces and vegetables of all kinds in air-conditioned plastic tunnels, with a cheap workforce from the Indian subcontinent. However, their fate is compromised by the ever-increasing use of desalinated water for irrigation. The overall disappointing results are due to a combination of factors. Firstly, the majority of landowners live in cities, and they do not feel involved in the process of production, so that about 40% of Liwa's agricultural areas were unproductive in 2013. In most cases, the agricultural income is considered a minor side-revenue compared to the main urban activities. Thus, the absentee landowners delegate the maintenance of their plots to foreign farm workers (mostly Pakistanis, Indians and Bangladeshis). In this context, agricultural activity is often seen as recreational agriculture, towards which the urban dwellers turn at weekends. The short stay of urban families in Liwa is one of the last legacies of a bygone Bedouin life. In this way, among the goals of agricultural development, the objective of settling a large-sized population in the rural hinterland was not fulfilled. Moreover, the failure of agriculture is due to its unsustainable nature. In Liwa, agricultural development is based on fossil water exploitation. Yet the development of cultivated areas and the free and unlimited use of water have led to massive waste and the exhaustion of underground water layers. This has resulted in the formation of a 50- to 100-km-wide depletion cone in the region of Liwa (Rizk and Alsharhan 2003). Therefore, the average depth of the wells has increased from 30 m in 1980 to more than 80 m today. Water degradation is also qualitative: the depletion of freshwater stocks has required the use of saltier deeper water. More than 65% of the wells produce water with a mineralisation ranging from 4000 to 14,000 mg/l, thus preventing the development of vegetable and fruit production without resorting to desalinated water. As a consequence, date palms and Rhodes grass fodder crops occupied almost 90% of Liwa's cultivated areas in 2009 as they show a good tolerance to salts. Faced with these economic setbacks and the overexploitation of the aquifer, the State now intends to preserve and reserve the exploitation of underground water stocks for the supply of Abu Dhabi, which simply means that the idea of a systematic development of productive agriculture must be abandoned in such arid latitudes.

4 Liwa, a Strategic Periphery Dedicated to the Metropolis

4.1 Priority and Vulnerability of the Urban Demand

In the United Arab Emirates, water is a scarce resource: in 2008, each resident had access to only 33 m^3 a year of renewable water, setting the country well below the threshold of water scarcity of 1000 m^3 per person per year. Abundance is still ensured as each inhabitant consumes on average 670–920 l of water per day, whereas the world average ranges from 160 to 220 l (EAD 2012). The paradox of this excessive consumption is due to the use of non-conventional water sources, mostly from fossil aquifers and desalination plants. The UAE are the world's third producers of desalinated water, behind Saudi Arabia and the USA. Thus, the entire volume of water consumed by the million residents of Abu Dhabi comes from only three desalination plants. Their total production is about 0.8 million cubic metres per day, providing a consumption of 560 m^3 per person per day. This makes the capital city highly vulnerable: its supply depends on the unique source of desalination plants, and its reserves are low. Resource volumes are 2 million cubic metres, that is to say hardly three days of storage. Such a dependence on the desalination infrastructure poses the challenge of water security in a global metropolis whose population will reach 3.1 million inhabitants in 2030. Should there be a natural disaster (hurricane, tsunami) or an industrial risk such as oil slick pollution, then the whole urban activity would be paralysed. Furthermore, there are many potential sources of conflict in the Gulf where oil stakes overlap terrorism and geopolitical rivalries between regional powers. The Gulf wars revealed that the desalination plants could be priority targets in the case of conflict. The UAE particularly fear their close neighbour Iran, with whom they are on bad terms. This is notably related to the unresolved issue of sovereignty over three islets (Abu Musa, Lesser and Greater Tunb), in the Straits of Hormuz, claimed by the UAE but occupied since 1971 by the Iranian army. The geopolitical stakes transcend this simple problem of sovereignty because they result more widely from the confrontation of two models of regional power, between the Sunni monarchies of the Gulf and the Islamic Republic of Shiite Iran. In this context of vulnerability, the government was forced to use the hinterland as a strategic storage space, able to sustain water consumption in urban populations in the case of political, ecological or social crises.

4.2 Liwa, a Strategic Reservoir of the Capital City Abu Dhabi

Based on the model of the strategic petroleum reserves decreed by the International Energy Agency, of which almost all the OEDC countries are members, the Emirate authorities decided to create a strategic water reservoir in the Liwa erg equal to 90 days of Abu Dhabi City's consumption (Fig. 4). This geographical spot was chosen for the large volume of sand in the Quaternary paleodunes, offering a high storage potential, as evidenced by the existence of the vast shallow fossil aquifer. The upper layer shelters a freshwater lens (up to 1500 mg/l) averaging 30 m in thickness and holding an estimated volume of 16 km^3 spread over 2400 km^2. The lower levels are more plentiful (up to 80 m in thickness), but their levels of salinity

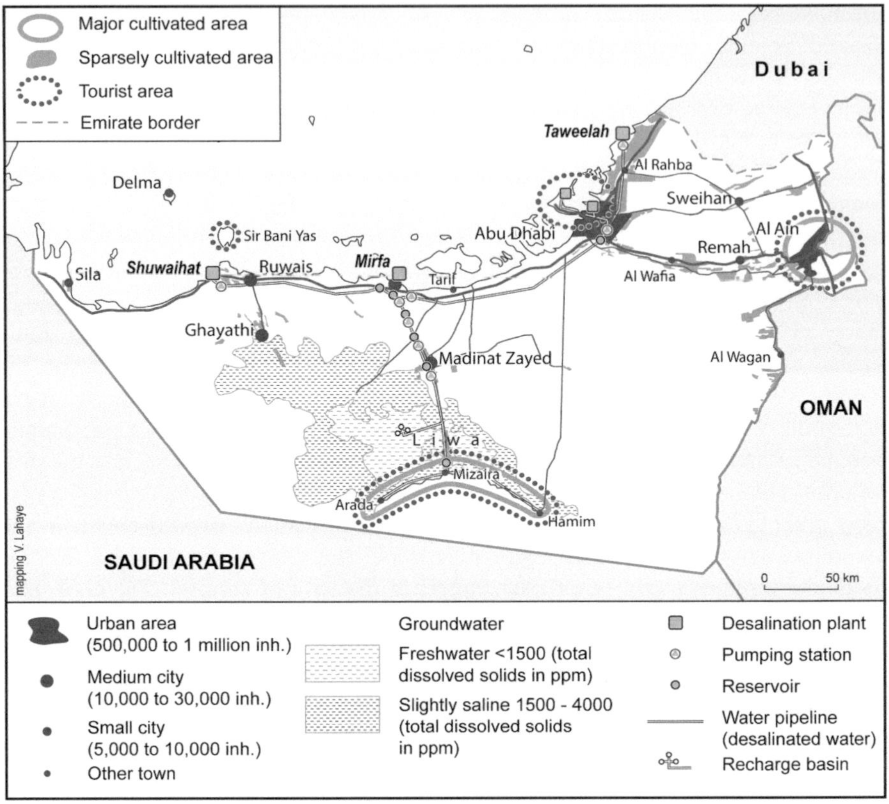

Fig. 4 Liwa, an oasis subordinate to Abu Dhabi

are high (Brook et al. 2006; Menche 2010). In this hydrogeological context, since late 2010, desalinated water has been injected from the coastal plants to Liwa through large pipes. Using the excess freshwater from production peak shaving in the desalination units, the recharge of the aquifer relies on three infiltration basins. Extending over 27 months, the recharge leads to a drinking water reserve of 26.1 million cubic meters (GTZ/Dornier Consult/ADNOC 2005; Menche 2010).

In the event of a crisis or a disruption of classic supplies from the desalination units, the groundwater reservoir would deliver 16.4 million cubic meters of drinking water to the capital city for 90 days, according to an input of 181,000 m^3/day, representing 182 l per capita per day. The choice of a groundwater reservoir stood out as the best economic and strategic solution. In fact, the construction of metal tanks on the ground would cover 250 ha and cost four times more than the underground option. To this would be added expensive maintenance and treatment costs to avoid water stagnation in the storage units: permanent recycling associated with disinfection and refreshing of water would be needed to keep it drinkable. Such costs and procedures do not arise in an underground reservoir, where the water quality stays naturally good over several years. It is clear that underground storage provides far higher security compared to surface containers, highly vulnerable to potential terrorist attacks or armed conflicts.

Such a world-first project and groundwater management are under close observation by the neighbouring states of the Gulf Cooperation Council, which are in the same situation of vulnerability and dependence on seawater desalination. Kuwait has a 5-day reserve, while Saudi Arabia, Qatar and Bahrain have only a 2–3-day reserve. This technical solution based on a "high-tech" water implementation is supposed to answer the water management issues experienced by all the Gulf metropolises. While the recharge of the Liwa aquifer is a strategy dedicated to urban populations, in the end it condemns productive agriculture in the area.

4.3 The Creation of a Patrimonial Agriculture

The implementation of a strategic drinking water reserve raises the challenge of resource-sharing and sector competition between rural and urban areas. In Liwa, the development of productive agriculture is henceforth incompatible with the new assignments of the region required by the Emirate authorities. With the current pumping rate for agriculture, the fossil aquifer would dry out within 50 years. Moreover, the increasing salinity and pollution by pesticides and fertilisers accelerate the degradation of water quality. Nitrates are already locally altering groundwater: in 1996, samples from Liwa wells revealed levels of nitrates ranging from 5 to 75 mg/l, while natural levels in groundwater range from 5 to 10 mg/l (ERWDA 2004).

After 30 years of unsuccessful attempts to achieve food self-sufficiency, the Emirate government has chosen to sacrifice agriculture and to run the race of water security. Since 2008, its water use policy has focused on the reduction in agricultural activity and strict control of resource use. This has resulted in the abandonment of wheat production and, since 2010, the prohibition of Rhodes grass fodder crops, which consumed 59% of agricultural water. In return, farmers of the Emirate who stopped growing fodder crops for dromedaries and small livestock received a 24.64-million dollar grant. Palm groves, responsible for 34% of agricultural water consumption, had to be systematically modernised, with the generalised use of a drip irrigation network over all the oasian area. Agriculture is gradually becoming a heritage where date production has regained its central place thanks to its historical and landscape aesthetic values. Although palm groves are altered and sometimes completely artificial products, they are invested with a high symbolic value. In comfortable luxury villas, the city dwellers are looking for their oasian origins and ancestral values of a Bedouin past, both recent on the scale of a few decades and distant regarding the radical socio-economic transformations undergone by populations in the area. The increase in weekend migrations, facilitated by the construction of asphalt highways, tends to re-create a link between the urban centres and the oasian hinterland, regarded as the space for identity reconstruction. As a new recreational resort periphery for nationals, Liwa's area is also strategically turned towards the development of international tourism. In the scenery of the great erg, the cluster of oases is afforded a patrimonial value, enhanced by the careful restoration of ancient forts and the establishment of very high standard hotels. The annual date palm festival is a major tourist event for nationals as well as foreign tourists. From now on, the luxury tourist paradise of the oasian arc is a complementary destination to the beach resorts and global cities of Dubai and Abu Dhabi.

5 Conclusion

With the creation of the first strategic groundwater reserve, the United Arab Emirates have launched a new form of desert area management, arising alongside the exploitation of hydrocarbons. The production of non-conventional water, as well as the use of renewable water resources, is a powerful vector of territorial organisation and transformation, as is the case of the spectacular reversal of Emirate space use. The need for water security revealed by the emergence of a global city has led to the eclipse of former concepts of the oasis as an agrosystem and traditional space organiser. In 1970, an expert in desert environments already pointed out that "from a strictly economic point of view, there is absolutely no doubt: agriculture will have to be eliminated from arid areas" (De Planhol and Rognon 1970). The UAE and the neighbouring states of the Gulf have headed in that direction after many decades of trying "to make the desert bloom." (Elhadj 2004). The current trend is towards "externalising" agricultural production by purchasing or renting lands in Africa or Asia, so that the oasian space is inevitably forced to evolve towards new development models. From now on, new assignments will have to combine tertiary sector issues with the need to invent a new territory invested with identity values suffering from the effects of accelerated globalisation. More than ever, the existence of the oasian cluster is entirely dedicated to and dependent on dominant coastal cities, challenging the changing oasian hinterland to look for a new stability.

References

Brook M, Houqani H, Darawsha T, Alawneh M, Achary S (2006) Groundwater resources: development and management in the Emirate of Abu Dhabi, United Arab Emirates. In: Mohamed A (ed) Arid land hydrogeology: in search of a solution to a threatened resource. Taylor and Francis Group, London, pp 15–35

Cadène P, Dumortier B (2011) Atlas des pays du Golfe. PUPS, Paris

Cordes R, Scholz F (1980) Bedouins, wealth and change. A study of rural development in the United Arab Emirates and the Sultanate of Oman. The United Nations University, Tokyo

Côte M (2002) Des oasis aux zones de mise en valeur: l'étonnant renouveau de l'agriculture saharienne. Méditerranée 3–4:5–14

EAD (Environment Agency-Abu Dhabi) (2012) Advancing sustainable groundwater management in Abu Dhabi. Annual policy, Abu Dhabi

Elhadj E (2004) Camels don't fly, desert don't bloom: an assessment of Saudi Arabia's experiment in desert agriculture. Occasional paper 48. Water Issues Study Group, School of Oriental and African Studies, King's College London, London

ERWDA (Environmental Research and Wildlife Development Agency) (2004) Assessment of the water situation in the western region of Abu Dhabi Emirate. Abu Dhabi

GTZ/Dornier Consult/ADNOC (2005) Combined artificial recharge and utilization of the groundwater resource in the Greater Liwa area. Pilot project final technical report

Heard-Bey F (1997) The tribal society of the UAE and its traditional economy. In: Al Abid I, Hellyer P (eds) United Arab Emirates. A new perspective. Trident Press, London, pp 98–116

Menche P (2010) Strategic water storage/recovery project in Liwa. Communication presented at the water day conference, Abu Dhabi, 30 June 2010

National Atlas of the United Arab Emirates (1993) United Arab Emirates University, Abu Dhabi

De Planhol X, Rognon P (1970) Les zones tropicales arides et subtropicales. Armand Colin, Paris

Rizk Z, Alsharhan A (2003) Water resources in the United Arab Emirates. In: Alsharhan A, Wood W (eds) Water resources perspectives: evaluation, management and policy. Elsevier Science, Amsterdam, pp 245–264

Sanlaville P (2000) Le Moyen-Orient arabe. Le milieu et l'homme. Armand Colin, Paris

Wilkinson J-C (1977) Water and tribal settlement in south-east Arabia. Clarendon Press, Oxford University Press, Oxford

Wood W, Imes J (2003) Dating of holocene groundwater recharge in the Rub al Khali of Abu Dhabi. In: Alsharhan A, Wood W (eds) Water resources perspectives: evaluation, management and policy. Elsevier Science, Amsterdam, pp 379–385

Wood W, Rizk Z, Alsharhan A (2003) Timing of recharge, and the origin, evolution and distribution of solutes in a hyper-arid aquifer system. In: Alsharhan A, Wood W (eds) Water resources perspectives: evaluation, management and policy. Elsevier Science, Amsterdam, pp 295–312

The Transformation of the Oases of Mendoza (Argentina): How the Provincial Socio-spatial Structure Was Reversed by the Crises of the 1980s and 2000s

Emilie Lavie, José A. Morábito and Santa E. Salatino

Abstract The oases of the Province of Mendoza, sited on the arid piedmont of the Argentinian Andes, have experienced some changes in their organisation, according to the civilisations that have cultivated these lands: Huarpes, Incas, Spanish colonisers and European migrants. Although these changes in societies transformed the agricultural orientation, the crises of the 1980s and 2000s did not radically change the typical Mediterranean landscape of such spaces. However, on the one hand they created a new type of landscape and, on the other hand, they had a deep impact on the socio-spatial and economic structures: former cores became secondary, and what used to be desert margins were developed until they became the new economic centre of the Province.

Keywords Globalisation impact · Agriculture expansion · Landscape changes · Socio-spatial structure

1 Introduction

Just like the *Ribeiras* of Cabo-Verde and the Oasis of Liwa in Abu Dhabi, presented in the two earlier chapters (Chapter "Insular Oases in Globalisation: the Ribeiras of the Cape Verde Archipelago, Fragmented and Fragile Areas on the Way to

E. Lavie
University Paris-Diderot, Sorbonne-Paris-Cité, UMR CNRS PRODIG, Paris, France

E. Lavie (✉)
Université Paris-Diderot, 5 Rue T. Mann, CC7001, 75205 Paris Cedex 13, France
e-mail: emilie.lavie@univ-paris-diderot.fr

J.A. Morábito · S.E. Salatino
Andean Regional Centre, National Institute of Water, Mendoza, Argentina

J.A. Morábito
National University of Cuyo, Mendoza, Argentina

E. Lavie and A. Marshall (eds.), *Oases and Globalisation*,
Springer Geography, DOI 10.1007/978-3-319-50749-1_13

Marginalisation" by Alexandre and Chapter "Liwa: the mutation of an agricultural oasis into a strategic reserve dedicated to a secure water supply for Abu Dhabi" by Cariou), the oases of the Province of Mendoza, Argentina, have experienced some changes in their spatial structures, due to their sudden entry into a new globalised economy, here clearly illustrated by wine market developments. This chapter, written by a geographer and two agronomists, aims to show how these changes have transformed small oases into globalised areas, well-known worldwide, in only one decade.

The oases of the Province of Mendoza, sited on the South-American Dry Diagonal, were created by diverting Andean rivers. In this arid land with 200-350 mm of precipitation a year, no rain-fed agriculture is possible. The snow-glacier regime of four streams enabled the creation and development of three big oases and two smaller ones (Fig. 1): starting from the North, the Mendoza River supplies more than half of the Northern Oasis; the Tunuyán River drains the Valle de Uco Oasis upstream, and the other part of the Northern Oasis downstream. Both the Northern and Valle de Uco Oases also use boreholes. Then, the Diamante and Atuel Rivers supply the Southern Oasis[1]. As shown in Fig. 1, these oases also have urban settlements, such as the Metropolitan Area of Mendoza, with more than one million inhabitants (INDEC 2010). In fact, 97% of the provincial population lives inside the irrigated areas (Ibid), making oases the real provincial territories.

Since the native period, these oases have seen some changes in agricultural orientation, until some of them became speculative areas cultivated with vineyards and orchards for the international market. Although these transformations in their economic vocation have changed the oasian landscapes throughout history, the Mendoza Oasis has always been the main core of the provincial socio-spatial structure. In fact, the sudden entry into economic globalisation in the early 1990s after the 1980s wine market crisis, illustrated by significant investment in modern irrigation by Northern Countries, widely transformed both the landscapes and socio-spatial structures. The aftermaths of a decade of liberal economy in the 1990s and a dozen years of national crisis (from 2002) have strengthened the development of a two-sided oasian system: on the one hand, a "traditional"[2] system, in which the older core has become marginal, and on the other hand, a former desert that has become the new economic centre of the Province. This chapter will first present a synthesis of the historic evolution of the oases and the changes induced in these last two decades in order to propose a schematic synthesis of the evolution of the system. It is based on (i) a bibliographical review, providing a state of the art of the

[1]This chapter will not deal with two smaller mountain oases in this Province: Uspallata (see Chapter "Mountain-Oases Faced with New Roads: Case Studies from the Andes and the Himalayas" by Lavie and Fort) and Malargüe.

[2]By "traditional", we mean the ninetieth-century system developed by European migrants. Agronomists use the term "traditional" to refer to the period 1880–1970.

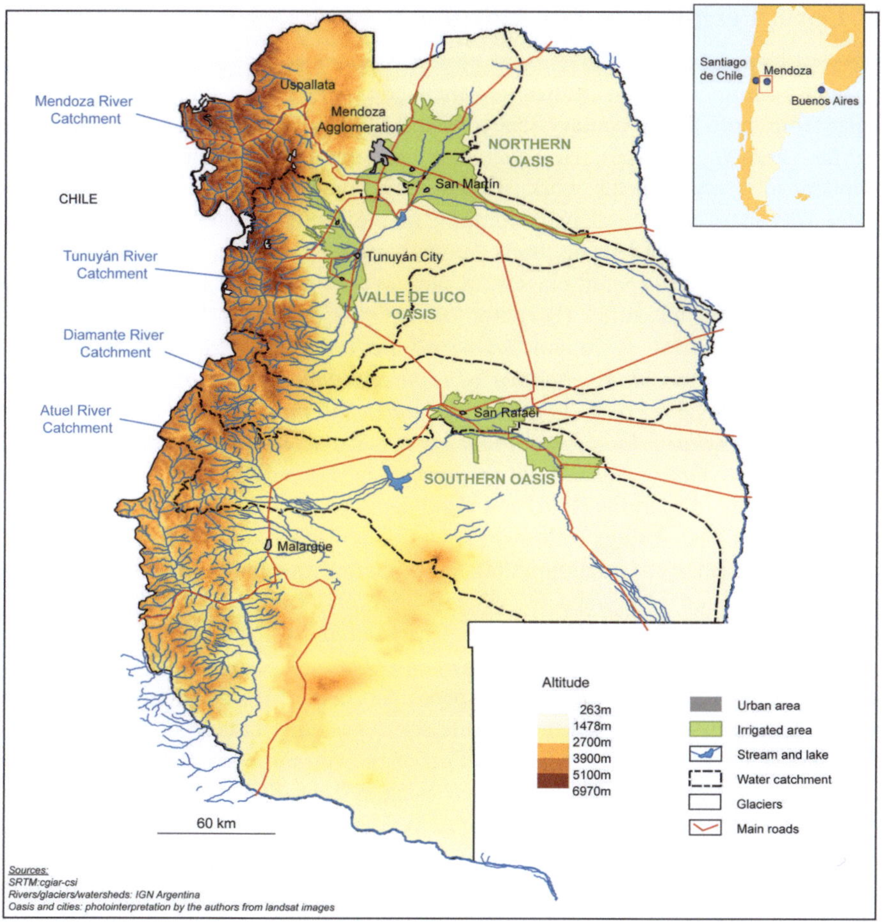

Fig. 1 The oases within the boundaries of the Province of Mendoza

historical evolution and general knowledge on the topic; (ii) personal fieldwork[3] involving observations, and interviews with farmers, local policy-makers, urban planners and researchers[4]; and (iii) an analysis of spatial dynamics from satellite imagery.

[3]The authors have carried out research on Mendoza's agriculture for some decades. Fieldwork on irrigation was shared between the three authors in 2006, 2008, 2012 and 2013. In 2014, the funding of the project "*Mendoza's oasian margins*" by the ComUE Sorbonne-Paris-Cité gave us the opportunity to add one month of fieldwork to interview researchers and farmers about the urban sprawl towards the green belt of the Metropolitan Area of Mendoza.

[4]During our 2014 fieldwork, we met specialists in urban sprawl in Mendoza: Mariana Sanmartino and José Reta (architect and geopolitician, Mendoza Land Use Planning Agency), Maria Teresa

2 Core, Peripheries and Margins

The relationships between centres and their peripheries have interested social scientists from more than a century. They were the topic of theorists of imperialism, of K. Marx and W. Sombart. However, the contemporary diffusion of the centre–periphery idea was first the work of theorists of inequality (e.g. Amins), then the model was later developed in geography in the 1980s (Grataloup 2004; Cattan 2006). Indeed, Reynaud in his book *Société, espace, justice* (1981) demonstrates the existence of socio-spatial classes, as well as an opposition between places that control and those that suffer (Brenetot 2004; Grataloup 2004). In 1992, Reynaud specified his definition again and "*shows all the ambiguity and the richness of the model Center ('where things happen')-Periphery (dominated, forsaken, integrated and exploited or annexed), to which he adds 'at the margins of centers and peripheries... isolated lands and blind spots'. These two territorial subsets are considered as dynamic elements of the periphery*" (Prost 2004).

The concepts of margins (Ibid) and blind spots add a certain discontinuity to this model made of nested rings following a gradient from the dominating centre to the dominated periphery (Grataloup 2004). In fact, they maintain the idea of an opposition between places/spaces, whatever the chosen spatial scale. Yet, spaces are dynamic and can evolve over time. The interactions, the mutually dependent links between asymmetric places, the flow of people and capital make these places self-regulating systems. Hence, some spaces can benefit from the situation, sometimes resulting in polarity reversal, while some systems remain identical and uniform and others represent a transformation of the former system (Ibid).

It is precisely these dynamics of the oasian system, that is to say the manifestations of these mutations between cores, peripheries and blind spots, which we seek to observe in the case of the oases of the Province of Mendoza. Our main hypothesis is that the entry flows of foreign assets have stimulated new flows of techniques (mainly for irrigation) and contributed to these overall mutations of the oasian landscape.

3 Historic Overview of Oasian Landscape Evolution

In a descriptive way, this first part seeks to understand the consecutive historical contexts that have contributed to create what is called the "traditional" Mendocinian landscape.

(Footnote 4 continued)

Vanderbock and Daniel Pizzolato (agronomists, National Institute of Agronomic Works, INTA), Maria-Elina Gudiño (geographer, Centre of Cartography and Formation in Land Use Planning (CIFOT) of the National University of Cuyo). We also had some interviews in the Municipality of Guaymallén but decided not to provide names for electoral reasons.

3.1 *Native Civilisations*

According to some data recorded at the beginning of colonisation, the Huarpes, a native community, developed agriculture on the alluvial cone of the Mendoza River from 1700 BC (Ponte 2006). In small groups of 100–150 people, they cultivated vegetables via a primitive irrigation system, diverting the Mendoza River using a geological fault (Chambouleyron 2004). It seems that in 1481, the Incas, coming from the North, arrived in the Cuyo region[5] and merged with the Huarpes, improving the primitive irrigation system:

> When the Spanish arrived, this site already had four large canals providing the local population with water[6] (Ibid).

3.2 *From the Spanish Colonisation to Independent Argentina*

Only 80 years after the Incas, in 1561, the Spanish Colonisers arrived in Cuyo and settled in the small village, later called Mendoza. They dug channels for domestic and agricultural uses: during the seventeenth century, the Mendoza Oasis (now part of the Northern Oasis) grew from 50 to 160 ha, organised in areas according to the network. The choices were to concentrate the production into two main sectors: cereals to be ground by hydraulic mills sited on channels and forage cultivated for cattle breeding. Thus, animals first came from the Pampa region and stayed in Mendoza for some weeks to be fattened before crossing the Andes Mountains to be sold in Chile. This was especially true for the Valle de Uco Oasis: at that time, it was closer to Chile, with densest hydrographic system and cheaper hydraulic structures due to a steeper slope, which favoured better pastures. Although in the Northern Oasis the average estate was 20 ha, in the Valle de Uco, owners had from 25 to 55 ha (Richard-Jorba 2006). Secondly, the oasis started to be cultivated with plants for alcohol and wine. During all the Hispanic colonisation period, grand families used Mendoza's oases as crossing points and pastures on the Pan-American road, based on farming, a stockbreeding pastureland system and landscapes. At the beginning of the ninetieth century, in the early independent Republic of Argentina[7], the need for hydraulic force in mills led to the development of new channels, enabling the expansion of cultivated areas (Montaña 2007).

> All channels "built during the 19th century, diverting main rivers, were the result of personal and private effort from farmers of that time. In this enterprise, the State did not intervene" (Chambouleyron 2004).

---

[5]*Cuyùm puùlli* in the Huarpe language means *Sand land.*

[6]All translations from Spanish or French were made by the authors.

[7]Argentina proclaimed its independence from Spain in 1810; it was definitively acquired in 1816.

3.3 "Los Mexicanos descienden de los Aztecas, los Peruanos de los Incas, y los Argentinos de los barcos"[8]: *Immigration as the Basis of the "Traditional" Mendocinian Landscape*

In the nineteenth century, railways were built all over the country to connect provincial cities with Buenos Aires for a better management. Being on the way to Chile, Mendoza City was connected to Buenos Aires in 1884 and to the Valle de Uco in 1911 (Chambouleyron 2004; Ponte 2006; Robillard 2010). In the early twentieth century, the colonisation of the south of the country brought railways to San Rafael in 1903 and to Colonia Alvear in 1912. Both were linked to Buenos Aires before being connected to Mendoza (Martinez Jurczyszyn 2011).

Most Argentinians are descendants of immigrants arriving from the late nineteenth century to the 1930s from European countries with wine-growing traditions, such as Spain, Italy and France. At the same time as the reduction in the breeding system, most profitable at that time in the Wet Pampa in eastern Argentina, a demand from the local market for low-quality wines developed (Robillard 2010). The livestock-farming/cereal system was replaced by a Mediterranean one. In parallel, the local industry changed from mills to wineries and canning factories (Montaña 2007); products formerly imported from the USA and Europe started to be produced locally in Mendoza (Romagnoli 2007). The establishment of the regional bank *Provincia* in 1888 enabled the development of the irrigation system and helped farmers to buy land, creating a new category of smaller (<5 ha) -and medium-sized (up to 30 ha) producers.

> Finally, taking the 1853 National Constitution as a basis, Mendoza acquired its first [Provincial] Water Law in 1884 and, from this year on, more than 52 special laws concerned irrigation (Chambouleyron 2005).

Since the late nineteenth century, the entire landscape has been changed: close to the urban settlements, a vegetable belt uses spring groundwater. Guaymallén Municipality in the Northern Oasis and Tunuyán Municipality in the Valle de Uco are representatives of this first green belt (*cinturón Verde*) and are still the main vegetable production zones, although the expansion of the agglomerations of Mendoza and Tunuyán is now problematic for such areas.

The second and main crop belt is occupied by permanent Mediterranean plants: vineyards, olive trees and orchards. Generally, olive trees are used as a belt for orchards or are planted in between vine rows. While the Northern Oasis grows stone fruits (peaches and plums), the Valle de Uco and Southern Oases have pome fruits, such as apples and pears (Chambouleyron et al. 2002). A century later, the Northern Oasis has a surface area close to 2500 km^2, the Valle de Uco more than 800 km^2 and the Southern Oasis around 1600 km^2 (DGI 1997).

---

[8]"Mexicans come from Aztecs, Peruvians from Incas, Argentinians from boats", a popular idiom.

Hence, in this Province,[9] oases produce fruits, olives and vegetables consumed as fresh products by the local market or canned in the industrial zones of the cities of Mendoza, San Rafael, Tunuyán and San Martín. Although first designated for the national market, these cans are now exported to other countries of MERCOSUR and associates.[10] Grapes are not consumed as fruit but vinified in wineries in industrial areas or within the cultivated perimeter.

Mendocinian viticulture experienced some changes at the end of the twentieth century: overproduction, and the decline in purchasing power and in consumption brought about an internal crisis (Bunel and Prévôt-Schapira 1994; Tulet and Bustos 2005; Robillard 2009, 2010).

> According to Robillard (2009), "to get out of the crisis, solutions diverge: while national politicians recommend stopping the monoculture, Mendoza's [provincial] government, concerned about protecting small producers, chooses a reconversion to both fruit and vegetable production".

These policies were not so divergent: while the national government suggested halting monocultures, provincial managers promoted the transformation of vineyards. Moreover, the transition from Spanish *estancia* to *huerta* with immigration waves was not as caricatural as presented: in fact, several wine crises obliged farmers to diversify their production to orchards and, while the Northern Oasis favoured vineyards, the main production in the Valle de Uco until the 1990s was fruit trees (apples and pears), cereals, alfalfa and potatoes. Indeed, the opening of a new road to Santiago de Chile directly from Mendoza City (see Chapter "Mountain-Oases Faced with New Roads: Case Studies from the Andes and the Himalayas" by Lavie and Fort) deprived the Valle de Uco of this traditional transit. Being far from the new route, producing wine was not as profitable as in the Northern Oasis.

Regarding the Southern Oasis, with its late connection to Mendoza and Buenos Aires, it was developed in the twentieth century. From the beginning, so as to avoid the mistakes made in Mendoza during the Hispanic colonisation, institutions were created to allow investment in land (agriculture and cattle), transport and industries. Organising the distribution of lands and a new irrigation system, they avoided *latifundios*. This enabled the creation of two oases, now connected: San Rafael linked to Mendoza, and General Alvear linked to Buenos Aires (Fig. 1). The Southern Oasis is mainly cultivated with orchards, olive trees and vineyards, mostly white grapes. Contrary to the other oases, horticulture is quite weak, but pastures and forests contribute to 25% of the total (DGI 1997). Due to its natural resources,

---

[9] The Province of Mendoza is one of three in the Cuyo regions. Argentina, as a federal country, established provinces at the end of the War of Independence.

[10] MERCOSUR: *Mercado Commún del Sur*, or Southern Common Market. The Southern American subregional bloc is led by Brazil and Argentina, but also includes Paraguay, Uruguay and Venezuela. Partly inspired by the European Union (preferential trade system, facility of migration, etc.), MERCOSUR is also associated with Chile, Bolivia, Colombia, Ecuador and Peru.

General Alvear has 350 mm/year of rain, which gives more opportunities for raising cattle and forest.

Finally, the three oases of the Province of Mendoza, with their economic model based on the coupling of Mediterranean cultures and agri-food-related industries, experienced some evolutions during the twentieth century, but no real structural change. However, the socio-spatial structures and the traditional Mediterranean landscape, inherited from the immigration waves of the 1890s, have been deeply affected by two crises: the 1980s Mendocinian and the 2000s Argentinian crises.

4 Economic Crises of the 1980s and 2000s and the Westward Expansion of the Oases

4.1 From an Overproduction Crisis to a Sudden Entry into Globalisation

Already in the 1930s, a local overproduction crisis had shocked the economy. The government responded by regulating the market, buying some wineries and encouraging farmers to diversify their production. However, in the late 1970s, wine consumption in Argentina experienced a substantial fall: each Argentinian consumed 90 L/year in 1970, 60 L in 1986 and 54 L in the early 1990s (*Vitivinifera* 2002, p. 17, cited by Tulet and Bustos 2005). As exports were quite negligible, a third of the wine production could not find any market.[11]

Unlike the earlier viticulture crisis, this one took place at a special political time: the beginning of a new democracy (from 1983). As the Northern and Central Oases saw an important expansion during the period of 1940–1978, they had no option but to pull the vines out.

> Yet, the Federal State "undertook to open the Province to globalisation and to its capital, to finance the reorientation of its vineyards to a qualitative one, the only alternative to get out of the crisis. Indeed, 1500 million dollars was invested in vitiviniculture during the 1990s decade" (Robillard 2010, citing Merino 2001).

The new orientation of the agricultural sector and the entry of globalised actors into the region were helped by two local and national contexts: firstly, a new economic orientation at the national scale. In fact, the new Republic was incapable of sustaining the economy in the 1980s. In return for the cancellation of part of the debt and the staggering of the remaining debt, the new President Menem (1989–1999) agreed to follow the Washington Consensus, such as the liberalisation of external trade and market deregulation, imposed by the International Financial Institutions (International Monetary Fund, World Bank). Then, in the Province of

[11]For the reader interested in Mendoza's wine history, see Romagnoli 2004, 2007; Richard-Jorba 2004; Tulet and Bustos 2005; Robillard 2009, 2010; Blanchy 2014.

Mendoza, local Governor Bordón started a provincial project of territorial marketing, so as to attract foreign investors (Velut 2002; Robillard 2010).

The result was the expansion of the oases, not towards the eastern desert as before, but to the piedmont: these places offered a better daily thermal amplitude, limiting diseases and favouring maturation, better soils, and better water quality and quantity. This change in the location of cultivated lands was helped by the presence of a permissive law concerning groundwater and new pressurised irrigation techniques.

4.2 Is Access to Groundwater a Discriminating Power?

The entry of the small oases of Mendoza into globalisation has been both economic and technical: global actors coming from Northern Countries and investing in the region have enabled the development of irrigation thanks to the adoption of a new technical system.

The new irrigated perimeters sited on the former piedmont desert of Luján (in the Northern Oasis) and the Valle de Uco took advantage of using groundwater (Fig. 2). Thus, they are independent of the superficial water variations and of the management choices as there is no quota for the groundwater extracted, only for the number of wells. The problem is that they also take water formerly used in the Northern Oasis:

1. Groundwater in the Valle de Uco resurges in the downstream valley by way of small streams merging with the Tunuyán River: this used to supply the Northern Oasis in superficial waters. In fact, taking groundwater upstream can deprive the downstream farmers of their historical resource (Fig. 2). A rise in the salinity of the Tunuyán River waters, used for irrigation, has been observed downstream, constraining owners to give up their lands or to change the type of crops (Chambouleyron et al. 2002; Morábito et al. 2012).
2. In the part of the Northern Oasis irrigated by the Mendoza River, we observed the same problem: water used upstream in Luján prevents farmers downstream in Lavalle or Guaymallén from using part of the discharge (Morábito et al. 2005; Lavie 2009; Barbier 2011).

   – Although the irrigation system was built by farmers downstream of both the Northern and Valle de Uco Oases, they are now dependent on the uses made by others, who arrived in the 1990s from outside the country. What used to be a piedmont desert has been converted into *la crème de la crème* of the provincial economy: a modern economy, advanced irrigation and agricultural technologies. Although some lower lands are abandoned daily due to water scarcity and sold for urbanisation, other farmers have found viable alternatives. This is the case of a farmer in San Carlos, Maria M., in the downstream green belt of the Valle de Uco, who used to produce garlic

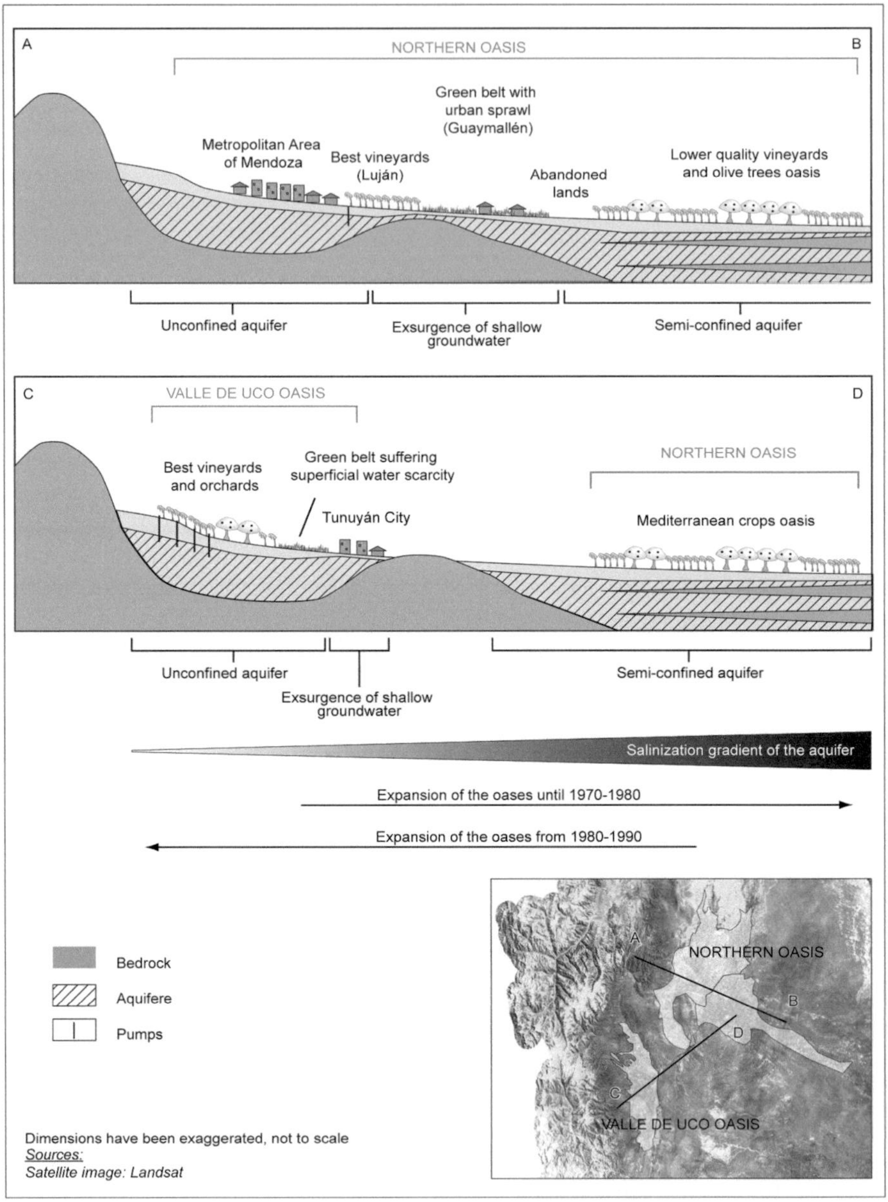

Fig. 2 Hydrogeological section and land uses in the Northern and Valle de Uco Oases

and tomatoes. However, after 5 years of receiving only 1–2 h of water every 8–10 days for 2 ha, she decided to turn her production into oregano. This is a hardier plant that suffers less from water stress than garlic or tomatoes. For this new crop, she is involved with other producers in a small cooperative that cuts, dries and packs oregano for the national market. Pedro S., a farmer in Colonia Molina, in the Northern Oasis green belt, also developed a new activity: he exchanged his unprofitable wine production for sweet potatoes destined for the local market. Thanks to the well his father dug in the 1970s when it was authorised, he has access to water of good quality and in the necessary quantity (contrary to the water received from the superficial network, contaminated by industrial effluents and supplied in a smaller quantity each year). He also belongs to a local cooperative.

These small farmers are typical of the nineteenth-century system of family workers in Mendocinian agriculture: in fact, they are all descendants of the late-comer immigrants who could only have the remaining lands, or more recent migrants from South America, mainly Bolivia. When the immigration waves from Europe started to decrease after World War II, the Province saw a new immigration wave from bordering countries in the 1960s–1970s (Cozzani de Palmada 2000).

These migrations can be explained by the characteristics of this regional economy, requiring a large workforce during harvest periods. Even when the attractive regional model was changed by the alteration of the exchange rate, these contingents of workers still entered the region. Many of them settled in Mendoza. Those who settled in rural areas mainly worked in horticulture and a few of them in the vineyards. In this context, although their imprint is less visible than that of former overseas migrants, they also acted on the territory (Montaña 2007).

In conclusion, and to answer the question "*Is access to groundwater a discriminating power?*", we agree with the hypothesis of Robillard (2009, 2010): the limiting factor for farmers is more the access to the capital needed to buy new irrigation technology rather than real access to water. Moreover, this access to technology has two main consequences: (1) the more you have access to it, the more important you are in the oasian system (ibid); (2) this technology has changed the oasian fine-scale landscape, as presented below.

4.3 A 15-Year Crisis

During the 1990s, liberal reforms (following the Washington Consensus) were fruitful in terms of economic growth. Argentina exported raw materials to finance its economy while, at the local level, the provinces spent on public employment. This chapter does not deal with the reasons for the crisis, which are quite difficult to simplify. To summarise, the country entered into a record indebtedness. The year 1998 marked the beginning of recession: in particular, the crisis started with the

40% devaluation of the *real*, as Brazil was the main trading partner. With peso–dollar parity, Menem's government sought to raise common awareness: there was no option but budgetary discipline. Yet, this policy was not supported by real investments for the modernisation of the national industry, or by structural reforms. In 2001, the country entered into a deep economic crisis: some banks closed, life savings were lost, inflation exploded and governments fell. The lower and middle classes suffered more than the rest.

In the aftermath of the 15-year crisis, in 2014 we observed a 70% gap between the official and the black market peso–dollar exchange rate, 30% inflation per year and still no investments.

4.4 A Rural Crisis Becoming Urban

One of the consequences of the crisis and some political choices is the loss of rural and industrial jobs. The brand "*industria argentina*" ("*Made in Argentina*") is not sufficient to reduce unemployment. The unemployment benefit policy is not an incentive: to be brief, finding a job is often synonymous with a loss of money and overtime hours are more taxed than ordinary ones.

Nowadays, the Metropolitan Area of Mendoza is an urban monster of 1.2 million inhabitants at the provincial scale without a priori planning: the city is growing due more to a rural exodus than to internal growth (Interview Land Use Planning Agency, fieldwork 2014). Hence, while at the beginning of the 2000s, the expansion was observed on the piedmont, since 2005, researchers have noted a new front in the green belt of the oasis. Today, urban margins are suffering from a phenomenon of urban front growth, and also urban sprawl. According to our interviews (see the Introduction), we believe this situation is due to three concomitant phenomena:

- An urban exodus by upper-class people, since the 1990s; they are fleeing from urban insecurity, created by the increase in the economic gap between upper and lower classes. They are also seeking a more natural way of life and do not want to live in the overpopulated Metropolitan Area of Mendoza. They settle in gated communities, here called *countries,*[12] with access to every urban services (such as drinking water, a sanitation system, electricity and waste management) thanks to the help of the political elite, who often live in such areas.
- a rural exodus of poor and unemployed people living in remote areas (e.g. General Alvear in the Southern Oasis); their living conditions are of such a low standard, they prefer to settle in the same areas as the *countries*, but with no access to all the urban services (less water, no waste and sanitation collection, no public transport, etc.).
- green belt urbanisation has been made possible thanks to the abandonment of land in the former core of the Northern Oasis. This is a hypothesis that we intend

[12] They use the English term even in everyday Spanish.

to work on in further fieldwork: Are the decrease in the irrigation superficial water available concomitant with the decrease in water quality (Lavie 2009; Barbier 2011), the difficulties in selling the production in a globalised market without investment, and the pressure of the property lobbies, some of the main explanations of land abandonment?

We can take as an example the Corralitos District in the Municipality of Guaymallén: there, the urbanisation level rose from 5 to 13% of the total area, between the years 2003 and 2013 (Alegre et al. 2014). This urbanisation is quite supported by the Municipality Intendant: in fact, the densification of a rural area can increase the local income and is rather interesting for electoral reasons. In the new Land Use Plan Law, the green belt has passed from a "rural" to a "rural with low density urbanisation" zone (interviews, fieldwork, 2014).

5 The Effect of the Crises on the Core–Periphery Pattern: New Landscapes, New Centralities

The core–periphery pattern proposed by Reynaud (1981) enabled the domination and mutual dependency links between close places to be observed. Grataloup (2004) and Cattan (2006) argued that this pattern was richer: they showed that an external effect could modify the inner system, sometimes resulting in centrality reversal. This was the subject of our analysis: on the one hand, through the entry of the oases into financial and technical globalisation, following local and national crises; on the other hand, by the synthesis of the historical evolution, looking back from the current situation at the successive mutations that the oasian system has experienced throughout history. Lastly, this review has enabled us to present this new landscape at two spatial scales: inside the oases and in between the oasian archipelago. Figure 3 is the synthesis of such an evolution.

In the aftermath of the endogenous crisis of the 1980s, the oases experienced the arrival of exogenous actors and techniques, investors from Northern Countries and pressurised irrigation (Montaña 2007; Robillard 2009, 2010). The permissiveness of local authorities created an agricultural front towards the West, developing a new production system that did not replace the old one but added to it, sometimes helping it to survive. On a positive note, the best part of the "traditional" farmers has taken advantage of the development of globalised agriculture in the piedmont: they can work as employees when they need cash, they sell grapes, rent land, etc. (Interviews, fieldwork, 2014). The older and downstream lands suffered a great deal from the 1980s crisis and new investors, stimulating and regenerating Mendocinian agriculture, have slowed down the lingering death of oasian viticulture. Others could not take this opportunity or, on the contrary, took another one: selling land for urbanisation, strengthened by a 15-year national crisis and a deep rural crisis.

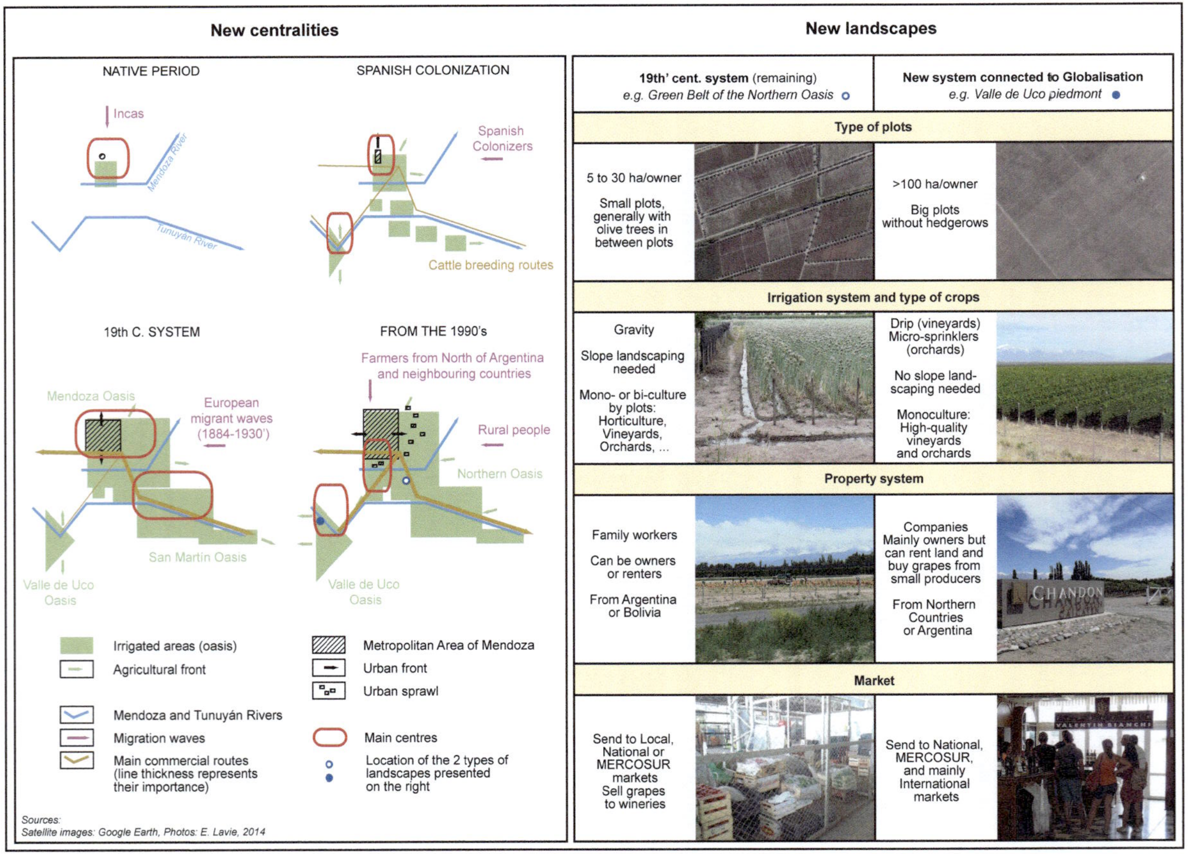

Fig. 3 New socio-spatial structures and landscapes in the Northern and Valle de Uco Oases

6 Conclusion

Just like Liwa (Chapter "Liwa: the mutation of an agricultural oasis into a strategic reserve dedicated to a secure water supply for Abu Dhabi" by Cariou), the Northern and Valle de Uco Oases show at the provincial scale a reversal of the former centrality model: while the Metropolitan Area of Mendoza is still considered a centre, it is no longer the only one. Furthermore, the Oasis of San Martín (part of the Northern Oases irrigated by the Tunuyán River), which used to be an agricultural nexus because of its location on the way to Buenos Aires, has been placed on the margins of the provincial production system, due to both rural exodus and water salinity. On the contrary, in the Valle de Uco Oasis, the formerly traditional farming area has become the new centre of globalisation.

Lastly, at the finest scale, inside the oases, the green belts are victims both of the power of the new globalised agriculture in access to water and of urbanisation. Incidentally, they can also be considered attractive places for urban people and are going to become future cores. The situation of the green belts of the Northern and Valle de Uco Oases and the position of the Southern Oasis in this panorama are the main questions of an ongoing research project named "oasian margins".

Acknowledgements The authors wish to thank Anaïs Marshall, co-editor of this book, for her participation in our latest fieldwork, Aldo Morsucchi, our driver during fieldwork interviews, and Malika Madelin for her valuable help with GIS.

References

Alegre F, Fernandez L, Ramirez M (2014) El ordenamiento territorial. Un instrumento de planificación y aporte a las estrategias de intervención territorial. XXXVII Argentinian Congress on Horticulture. Mendoza, Argentina, 25 Sept 2014

Barbier M (2011) La qualité du système d'irrigation agricole de l'oasis du río Mendoza, Argentine. Master thesis, University Paris-Diderot, France, p 67

Blanchy G (2014) Le vignoble argentin de Mendoza et l'influence française, XIX-XXIè siècle. Paris, l'Harmattan, Coll. Recherche Amériques Latines, p 193

Brenetot A (2004) Justice spatiale. Hypergeo (On-line publication)

Bunel J, Prévôt-Schapira MF (1994) Nouvelles politiques et entrepreneurs: le cas de la Province de Mendoza. In: Bunel J, Montero-Cassasus C Prévôt-Schapira MF (Eds) Les entrepreneurs sous le choc libéral. Le cas du Chili et de l'Argentine, Rapport MRT/Document de recherche du CREDAL, pp 62–201

Cattan N (2006) Centre-Périphérie. In: Ghorra-Gobin C Dictionnaire des mondialisations, Armand Colin, pp 47–49

Chambouleyron J (2004) El aprovechamiento del agua (On-line publication). www.irrigacion.gov.ar

Chambouleyron J (2005) Riego y Drenaje. Técnicas para el desarrollo de una agricultura regadía sustentable. Tomo II. EDIUNC, Mendoza, Argentina. ISBN 950-39-0176-6

Chambouleyron J, Salatino S, Drovandi A, Filippini M, Medina R, Zimmermann M, Nacif N, Dediol C, Camargo A, Campos S, Genovese D, Bustos R, Marre M, Antoniolli E (2002) Conflictos ambientales en tierras regadías. Evaluacíon de impactos en la cuenca del Río Tunuyán, Mendoza, Argentina, UNCuyo-Foncyt-INA-EON. Mendoza, Argentina, p 185

Cozzani de Palmada M (2000) Inmigrantes limítrofes en Argentina.¿Tolerancia o rechazo? *Les Cahiers Alhim* n°1

DGI—Departamento General de Irrigación (1997) Plan hídrico para la Provincia de Mendoza. Ésta es Mendoza. Mendoza, Argentina

Grataloup C (2004) Centre-périphérie. Hypergeo (On-line publication)

INDEC (2010) General Census of population. On line data: http://www.indec.gov.ar/nivel4_default.asp?id_tema_1=2&id_tema_2=41&id_tema_3=135

Lavie E (2009) Activités anthropiques et qualité de l'eau dans l'oasis de Mendoza (Argentine): diagnostic, enjeux et durabilité. Ph.D., Université Bordeaux III, France, p 385

Martinez Jurczyszyn C (2011). Historia de la Colonia Alvear (1901-1914). Revista de historia americana y argentina n° 46-2. Scielo versión On-line ISSN 2314-1549

Merino J (2001) Un ambiente de libertad económica logró el cambio. Fenomenal reconversión del viñedo argentino. Bodegas y Terruños, Mendoza XI: 8–11;

Montaña E (2007). Identidad regional y construcción del territorio en Mendoza (Argentina): memorias y olvidos estratégicos. Bulletin Français d'Études Andines, vol. 36, n°2: 277–297

Morábito JA, Mirábile C, Manzanera M, Cappe O, Tozzi D, Mastrantonio L (2005). Evolución de la salinidad de suelos regadíos e incultos en el área del río Mendoza. XXth national congress on waters (CONAGUA)—IIIrd symposium on Southern Cone's hydric resources, DGI-Mendoza, Argentina, vol 1, pp 1–10

Morábito JA, Salatino SE, Campos S, Maffei J, Mastrantonio L, Stocco A, Mestres Sanchis F, Genovese D (2012) Estimación de los incrementos de la salinidad del agua superficial en la cuenca del río Tunuyán como consecuencia del impacto climático. Congress on Irrigation. AERYD-CERYD, Albacete, Spain

Ponte R (2006) Historia del regadío. Las acequias de Mendoza, Argentina. Scripta Nova, Revista electrónica de geografía y ciencias sociales. Barcelona, Spain, vol X, nº218 (07)

Prost B (2004) Marge et dynamique territoriale. Géocarrefour 79(2):175–182

Reynaud A (1981) Société, espace et justice: inégalités régionales et justice sociospatiale. PUF, Paris

Richard-Jorba R (2004) ¿Echar raíces o hacer la América? Un panorama de la inmigración europea hacia la región vitivinícola argentina y algunos itinerarios económicos en la provincia de Mendoza, 1850-1914. Les Cahiers ALHIM, Paris 8, Migrations en Argentine II, (9)

Richard-Jorba R (2006) Formación, crisis, y reorientaciones de la vitivinícultura en Mendoza San Juan, 1870-2000. Aportes para el estudio del sector en la Argentina. Boletín Geográfico, Neuquen, Argentina. n° 28:79–122

Robillard J (2009) Des glaciers au vignobles: gestion de l'eau et stratégies d'irrigation dans les terroirs vitivinicoles de la province de Mendoza. Master thesis, Université Paris I Panthéon Sorbonne, France, p 165

Robillard J (2010) L'accès à l'eau dans l'Oasis de Valle de Uco (Province de Mendoza, Argentine). Quand l'agriculture sort des périmètres irrigués traditionnels pour transformer le désert en vignoble. Master thesis, Université Paris I Panthéon Sorbonne, France, p 99

Romagnoli EP (2004) Inmigración europea y producción artesano industrial en Argentina: la metalurgia inducida por la vitivinicultura moderna en Mendoza (1885-1930). Les Cahiers ALHIM, Paris 8, Migrations en Argentine II, (9)

Romagnoli EP (2007) Immigrants métallurgistes français et vitiviniculture moderne à Mendoza et San Juan, Argentine (1885-1930). Les Cahiers d'Outre-Mer n°239:283–299

Tulet JC, Bustos R (2005) Le vignoble de Mendoza (Argentine), entre vin de table et vin de qualité. Les cahiers d'Outre-Mer n° 231–232:281–300

Velut S (2002) Argentine, géographie d'une crise. L'espace géographique 178(2):175–178

General Conclusion
The Oases Challenged by Glocalisation

Jean-Louis Chaléard, Sabine Planel and Thierry Ruf

The oasis is a well-studied subject in the scientific literature, which this book revisits in the light of the most recent changes in the modern world. The thirteen chapters assess the current state of play and highlight the current dynamics, sometimes contradictory, which affect oases. A wide range of cases is presented, as can be seen on the map in the introduction, enabling an analysis of the oasian phenomenon in all its complexity. The oases considered are found in North Africa and the Sahara (although there is no example from the southern Sahara in contact with sub-Saharan Africa), Latin America, and Central Asia as well as the Middle East and the Cape Verde islands.

The geographical approach is favoured due to the scientific origin of the authors, which is demonstrated by an important reflection on the landscapes and the changes in the territories. However, the varied backgrounds of the contributors enable other points of view and other themes to be taken into account, based on society, history, and literature.

One of the lessons of this book is that the oasis, an apparently simple subject, is striking in its diversity, posing a formidable problem of definition behind the simplistic images that are sometimes conveyed. These questions of definition are first addressed before focusing on the heart of the matter; the changes linked to

J.-L. Chaléard
University Paris 1 - Panthéon-Sorbonne, UMR PRODIG, Paris France
e-mail: chaleard@univ-paris1.fr

J.-L. Chaléard
PRODIG, 2 rue Valette, 75005 Paris France

S. Planel
Institute of Research on Development, UMR PRODIG, Paris France

T. Ruf
Institute of Research on Development, UMR GRED, Montpellier France

E. Lavie and A. Marshall (eds.), *Oases and Globalisation*,
Springer Geography, DOI 10.1007/978-3-319-50749-1

globalisation and the actors in these developments who enable the local to be connected to the global.

Understanding the Concept

Water and Society

The first issue is the definition of the oasis, its limitations, and even the relevance of the term itself. Although this is explicit in some texts, particularly in the first part of this book, it appears in almost all of them.

The "classic" image of the oasis goes back, in French, to the descriptions of the geographer Jean Brunhes at the beginning of the twentieth century or to those of André Gide in *Les Nourritures terrestre*, and still to today's tourist brochures. The oasis is an irrigated area with terraced crops and intensive techniques, which is often thought of as an "island" in the desert, a very ancient Épinal image as F. Alexandre recalls in his text on the Cape Verde oases. What remains of these images?

All the texts agree about the importance of water. Thus, B. Veyrac-Ben Ahmed and S. Abdedayem, who discuss this concept, emphasise the special features of oases in an arid environment, in which water is the key to the spatial and social organisation. Similarly, most authors show that the reality is particularly complex. M. Kober, focusing on the literary texts of Siwa, provides one of the keys to the definition of oases. An "oasis" was initially a Copt term to describe places far from the Nile Valley. The literature shows that these were not just a special area linked to water. They were places that had escaped, at least partially, from power. The oases have a history and are often haunted by mysteries associated with the past. They also have other important functions. Stopping places on the great trade routes (as recalled by P. Cadène regarding Rajasthan) or ancient crossroads (like the oases of Peru studied by É. Mesclier, A. Marshall, C. Auquier, and J.L. Chaléard), the oases often play a vital role in trade. The oasis is thus not only a space organised around water and created by the presence of water, but also one that combines a variety of functions with an ancient life.

What is the relevance to other worlds of a concept forged in Africa? (This question is raised by É. Mesclier and co-authors regarding the oases of the Peruvian coast). Should the definition be enlarged to take into account the many different situations?

Oases

The chapters reveal the great diversity of examples. The oases, as anthropised spaces in arid or semi-arid conditions, are found in very varied situations. They occur in the mountains (e.g. in the Andes and the Himalayas), along some valleys with limited and intermittent water courses (such as the oases of the piedmont), in endorheic situations (like in Central Asia), and in depressions or coastal situations where very little freshwater is available (Fig. A.1). They differ in their size, their organisation, and their principal activities. The oases of Central Asia, or the piedmont oases of Peru, present a different organisation to those of the central Sahara. The Orange River Valley, almost 300 km long (described by D. Blanchon), is very different, in its size, the origin of its water, and the South African environment where it occurs, from the oases in the valley bottoms, isles of greenery in the Cape Verde islands, mistreated by the colonial powers, presented by F. Alexandre. The oasis is primarily a cultivated area, but those studied by P. Cadène, in Rajasthan, are essentially urban, having prospered for a very long time from long-distance trade.

Territories in Movement

In addition, the oases are not fixed places. Most of them are very old and have experienced many trials and tribulations during their history. The image of a stable territory, almost unchanging, an image largely inherited from colonisation, is challenged in several chapters. In the case of the oases of Southeast Arabia studied by J. Charbonnier, the oasian agrosystem is the result of a long history of the domestication of plants (including the date palm) and the establishment of watering systems, which have gradually shaped it over several millennia. The example of the Chicama Valley in Peru, analysed by C. Clément, highlights the variations in the area of oases between the eleventh and sixteenth centuries, alternating between expanding and shrinking, in relation to the socio-economic organisation that played a determinant role in these developments.

Examining the past leads to an understanding of the present. É. Lavie and M. Fort, studying the location of the oases of Uspallata in the Argentinian Andes and of Mustang in the Nepalese Himalayas, emphasise the age of their human settlement, dating to several millennia, and reveal the role of their accessibility and the existence of ancient roads in their growth.

Thus, the current changes are part of a series of developments that have affected oases over the course of history. This is clearly shown by those in the province of Mendoza in Argentina (cf. É. Lavie, J. Morábito, and S. Salatino), which have seen many changes, first with their exploitation by the Huarpes, followed by the Inca then the Spanish conquests, and then the arrival of European migrants in the nineteenth and twentieth centuries, before the recent disruptions linked to the crises of the 1980s and 2000s, which have led to a great expansion of vineyards in the new irrigated lands.

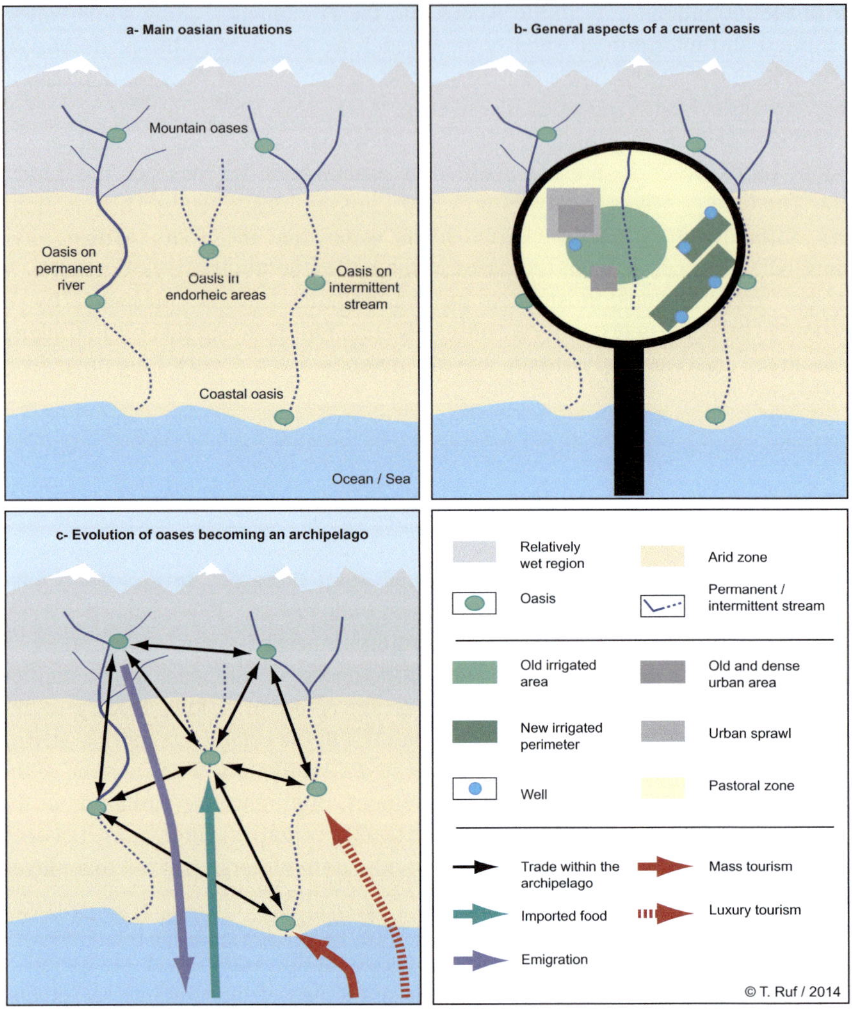

Fig. A.1 Spatial reconfiguration of oases in the Globalisation

While the development of oases is not linear, some characteristics are found at different periods: despotism, colonialism, confrontation with distant political powers, and times of opening up and of closure. In the oasian space and its periphery, rival groups often agree with or confront each other: for example, the nomad pastoral tribes and the groups of settled farmers in Africa or central Asia. The oasis is frequently depicted as a community living in a dense urban or village space where very specific categories coexist, varying according to each situation, such as warriors, traders, priests, artisans, farmers, and miners. Nevertheless, the powers are distributed in oases based on the control of different flows: water of course, people, raw and transformed materials, imported food, and money.

The Acceleration of Changes During Globalisation

Spatial Changes

Although the oasis has never been a stable space, due to the fragile balance between access to water, dominant social organisations, and conflicts within empires or between rival empires, this book highlights the force and the complexity of current changes, due to rapid developments, accompanied by phenomena of resilience. Differentiated dynamics are at work, especially within thc framework of globalisation, the central theme of the book. Clearly, the oases are unequally integrated into the current great exchange networks with the rest of the world, but the local and global forces in operation lead to significant developments everywhere, as a result of contradictory dynamics.

In spatial terms, there is an extension of cultivated zones, usually completely separate from the old land. Water has become accessible outside the framework of previous collective organisations that used to manage its distribution. It no longer only depends on the new structures made possible by drilling techniques and relatively cheap energy. Irrigated areas are being created in the peripheral, often pastoral, spaces (Fig. A.1b). The mobilisation of new water comes from various sources: wells in shallow or deep water tables (as in southern Tunisia studied by I. Carpentier and A. Gana), the construction of large dams (as analysed by D. Blanchon in the Orange River Valley), or even the diversion of water from streams at the cost of enormous works (e.g. in Olmos, in northern Peru, studied by É. Mesclier et al.).

This development is accompanied by setting up new crop systems related to worldwide demand, such as fruit crops for northern markets since the end of apartheid in the Orange River Valley. Sometimes, these changes began during colonisation (such as the *deglet nour* date plantations in southern Tunisia), but they have increased and expanded everywhere. The current changes intensify the internal diversity of oases with the appearance of new irrigated areas or the organisation of the landscape, the techniques employed, and the irrigation systems that differ from those of the older oases.

Moreover, the oases have not escaped the phenomenon of industrialisation, which has deeply transformed them, providing jobs and incomes and marking the landscapes: for example, the exploitation of hydrocarbons in many oases in North Africa and the Middle East, the industrial complex of Gabès.

Tourism is growing everywhere, despite unexpected events that can temporarily reduce the numbers of visitors (as in Tunisia, during the revolution of 2011). Based on the beauty of their landscapes and their rich architectural heritage, tourism is one of the major factors of the transformation of many oases. P. Cadène describes the conversion of the oases of Rajasthan into luxury tourist sites. Tourism may also change the role and function of agriculture, such as in Liwa (Abu Dhabi), where A. Carriou shows that it has developed a heritage function, with the palm groves becoming a tourist attraction.

At the same time, the oases are urbanising. The old towns are spreading into both the oasis and its periphery, over areas considered uncultivated and unsuitable (even though they in fact belong to the old oasian communities). New lifestyles are appearing, like in southern Tunisia where multiple activities are developing. The growing role of migration is an important and ambivalent phenomenon, integrating the populations into the rest of the world, but having strong negative consequences on agriculture with the departure of young workers and a change in income sources.

Crises and Competition Within Oases

These developments do not occur without clashes and changes in the organisation of territories. The new activities are not only a source of jobs, but also of competition and pressure on resources. In general, their growth is at the expense of the old activities. They are big consumers of water: this is true of tourism (cf. the oases of Jérid) and of industry (like in Gabès). The overall water balance is no longer equilibrated, and the water tables are exploited without any regulation, so that they become deeper each year, to the point of drying up the sources and devices of the old oases. The new boreholes for the new irrigated areas can lower the water tables (as in the oasis of Ica in Peru described by E. Mesclier et al.). In contrast, the accumulation of water supplies can, in some cases, lead to soil saturation (like in Viru, also in Peru). In addition, urbanisation competes with agriculture where space is at a premium and especially where water is rare (whether it is in Tozeur or Gabès in southern Tunisia). In extreme cases, like in Lima, urbanisation has led to the disappearance of most of the previous agriculture.

The oasian crisis takes many forms, as shown by the example of the Cape Verde oases analysed by F. Alexandre: spaces marginalised by the issues of reorganisation of global exchanges, the oases of this country have been faced with the monopoly of lands, the displacement of resources (water, work, biodiversity), and the abandonment of specific territories.

These developments deeply transform the organisation of the oasian space, not only by the appearance of new activities but also by the use of new agricultural techniques. Often, the irrigated agriculture of the older parts, at the centre of the oasis, is in decline, while the new areas of the periphery are growing. The dynamics of oases vary depending on their situation. In the province of Mendoza in Argentina (É. Lavie, J. Morábito, and S. Salatino), some parts are declining (like the oasis of San Martin, marginalised by the rural exodus and the salination of lands) while the Uco Valley oriented towards vines and wine is booming. In Liwa, A. Carriou describes the inversion of the roles and powers between the inland oasis and the coastal urban big bang. Pushed to the limit, the oasis has become a pseudo-reservoir of desalinated water to safeguard the urban development.

Local and Globalised Actors

From Local to "Glocal"

The almost unanimous observation is that of a crisis originating from the pressure on resources, the marginalisation of actors and strong elements of revival, but unequally distributed and unequally important depending on the oases. This concerns local phenomena, which are also related to more global situations (Fig. A.1c). In the oases, as elsewhere, there are echoes of encompassing developments, national or worldwide. Thus, the competition between small farmers and agribusiness, although not specific to oases, particularly marks these spaces. Moreover, the oases suffer from the consequences of changing national and international political situations and may be stakeholders in them. This is true of the oases of Baglan, north of Kabul, affected by the fall of the monarchy, the war against the USSR and the arrival in power of the Taliban (H. Kreutzmann, S. Shütte). It is also the case in Tunisia where the consequences of the 2011 revolution are profound. The Orange River Valley was also affected by the end of apartheid, which changed the water distribution system. Today, the oases are evolving in the interplay of local and global phenomena and are still influenced by the intervention of actors who are outside the oasian framework.

Increasing Numbers of Actors

Many chapters are interested in the actors of contemporary transformations who differ in type and scale and who contribute in their own way to the insertion of oases in globalisation and/or to retaining their local features.

The small agricultural producers, old inhabitants of the oases, are still present, sometimes only in survival conditions. However, they continue the old agriculture of oases, which is one of their attractions to the extent that they may play a role in maintaining the Épinal images so attractive to international tourists. Some of these farmers participate in the most notable transformations by growing specialised crops for world or urban markets, benefiting from the opportunities provided by the growth of towns, as in southern Tunisia (I. Carpentier, A. Gana), integrating into the encompassing dynamics.

Although the old institutions, such as the rural communities in Peru, have some difficulty withstanding the new context created by the liberalisation of the water and land markets, some farmers' organisations may emerge or experience a boost, such as the cooperatives and many other institutions that have recently developed in southern Tunisia, participating in a new connection between the oases and the world.

However, the major factor is, without doubt, the increasingly powerful intervention of big national or international agribusiness companies, which monopolise the land and the water and are found in almost all the oases. These enterprises are generally connected to a globalised market into which they want to integrate the local economies. In addition, tourism is often in the hands of large operators outside the oasis. Nevertheless, even in the most dynamic sectors, the locals can play a determining role: in Rajasthan, the old Maharajahs have greatly contributed to the development of tourism in the oases by converting their palaces into luxury hotels (cf. P. Cadène).

The link between the local and global scales is not only made by the transnational investors. The international institutions play a major role, whether it is the FAO in agricultural development programmes or UNESCO in the protection and restoration of ancient architectural forms (e.g. in southern Tunisia). All contribute to imposing globalised standards. Like everywhere in the world, but with more visibility here than elsewhere, the norms of UNESCO, in sustainable development or territorial marketing, are spreading and gradually erasing the original special features of the oases. The architectural norms are becoming homogeneous and meet the requirements of the new users of these spaces, the tourists. The aesthetics are changing.

Lastly, the oases often engage in the spontaneous dynamics of opening up, via an increase in international migrations, which provide income that enables some to stay and which may be the result of various initiatives in agriculture and tourism, in search of those resources that seem to be lacking in their near environment.

What About the State?

The position of the State may appear paradoxical in some ways, both weak and strong. Present to varying degrees in the oasian hinterlands that make up the national territories, the State does not seem particularly concerned by oasian development, including in those national contexts characterised by the presence of large public developers (Tunisia for example). There is no trace of an "Oasis Plan" in the chapters of this book. In contrast to the developmental view that seeks to valorise the comparative benefits of each territory, the public authorities seem to have generally neglected the marked special features of these environments.

Yet, at the same time, the State is a particularly important actor to whom many authors refer concerning local development policies and beyond overall economic policies. Its action is not necessarily direct or specific to oases, but it contributes to reshaping them. The new Tunisian policy since the revolution, to promote the rural areas of central and southern Tunisia, falls into this category. As in many countries, the action of the State that was once very directive is now reduced, in the context of liberal measures introduced in the 1980s–1990s. In Cape Verde, after independence, F. Alexandre observes an intervention by the authorities to develop the oases, with land reform and the extension of irrigation. Today, the liberal orientation of many national economies has set up a new deal and opened up oases to the outside. Although the policies steer the forms of development, they are not the only ones. Thus, in the province of Mendoza in Argentina, the extension of the vineyards is allowed by a permissive legislation amoung groundwater (cf. É. Lavie, J. Morábito, and S. Salatino).

In many countries, the State plays a fundamental role in the organisation, direction, and financing of public infrastructure networks (especially roads) that greatly affect the fate and the eventual opening up of oases. Thus, the oases of the central Argentinian Andes and the Nepalese Himalayas have benefited from the creation of international roads (E. Lavie, M. Fort). In Rajasthan, the construction of roads by the government has favoured the rise in tourism, even though they initially had a military objective, due to the proximity of the border with Pakistan (P. Cadène). In South Africa, the role of the State is an old one: under colonisation, the exploitation of the Orange River Valley was already an integral part of the apartheid policy, with the extension of the oasis being of advantage to the Whites (D. Blanchon). In addition, the oases of Baglan experienced growth due to the action of the authorities in the twentieth century (H. Kreutzmann, S. Shütte).

In many regions, especially in North Africa, the State restructures the civil society and determines in advance the institutional tools of collective mobilisation; it directs priorities towards a local, participative, and territorial development; and it constitutes a new decentralised basis for cooperation and partnership. In short, it regulates everywhere. By the enactment of land policies, it determines the conditions of access to the ground. However, in the current context of liberalisation, it seems more like a facilitator of local or foreign initiatives, especially those of big companies, than a direct actor in the major changes in oasian spaces and societies.

Between Glocal and Globalised Upgrading
What Is the Existence of Oases?

All the contemporary transformations raise the question of the existence of the oasis, as a unique space, today. D. Blanchon emphasises in his text that the income of the producers in the Orange River Valley depends more on variations in world prices than on fluctuations in the water level regulated by the big dams. B. Veyrac-Ben Ahmed and S. Abdedaiem show that the current developments are completely transforming the image of the oasis, which is fading with the expansion movement, the substitution of drilling, and the multiplication of contradictions between the actors and in the management of water. They question whether one should talk of oases in southern Tunisia or rather spaces with certain characteristics of oases. É. Mesclier et al., considering the oasis as a crossroads and an interconnected system, underline the relevance of the concept for investigating this type of space, while highlighting the disappearance of its older characteristics, and its exemplary nature in rising to the challenges of globalisation.

Beyond the simple observation of the special features of the oasian space, this book also places the oases and their changes in the developmental trajectory that falls within encompassing phenomena, especially globalisation. However, to what extent is the oasis more an example of a situation, of a time, in a trajectory? A time built on a tension between the dynamics of opening up and of closure, of connection with the world, and of looking inwards. This situation is particularly sensitive to the phenomenon of glocalisation, to this direct contact between to ends of the scale, the local and the global.

How can this impact of glocalisation be understood? Does it have a particularly large effect on these places so archetypal of their immediate environment for finding the necessary resources far away? Either they can access the national resources, or there is a better valorisation at a smaller scale. Can one see in this search for contact with the world an effect of the oasian structure? It is as if their already very marked otherness in their closer relationships does not increase with distance (physical, symbolic, cultural) and that their recourse to the global (its norms and resources) occurs relatively easily, previously like today. Clearly, one of the contributions of this book is to consider this link between the special features of an environment and the fundamental movements of the modern world.

Index

E. Lavie and A. Marshall (eds.), *Oases and Globalisation*,
Springer Geography, DOI 10.1007/978-3-319-50749-1

MIX
Papier aus verantwortungsvollen Quellen
Paper from responsible sources
FSC® C105338
FSC
www.fsc.org

If you have any concerns about our products,
you can contact us on
ProductSafety@springernature.com

In case Publisher is established outside the EU,
the EU authorized representative is:
Springer Nature Customer Service Center GmbH
Europaplatz 3, 69115 Heidelberg, Germany

Printed by Libri Plureos GmbH
in Hamburg, Germany